# Buying Time for Climate Action

## Exploring Ways around Stumbling Blocks

# Exploring Complexity



For four centuries our sciences have progressed by looking at its objects of study in a reductionist manner. In contrast complexity science, that has been evolving during the last 30–40 years, seeks to look at its objects of study from the bottom up, seeing them as systems of interacting elements that form, change, and evolve over time. Complexity therefore is not so much a subject of research as a way of looking at systems. It is inherently interdisciplinary, meaning that it gets its problems from the real non-disciplinary world and its energy and ideas from all fields of science, at the same time affecting each of these fields.

The purpose of this series on complexity science is to provide insights in the development of the science and its applications, the contexts within which it evolved and evolves, the main players in the field and the influence it has on other sciences.

---



For the complete list of volumes in this series, please visit www.worldscientific.com/series/ec

**Exploring Complexity – Volume 8**

# Buying Time for Climate Action

## Exploring Ways around Stumbling Blocks

Editors

Jan Wouter Vasbinder
*Para Limes, The Netherlands*

Jonathan Y H Sim
*National University of Singapore, Singapore*

**World Scientific**

NEW JERSEY · LONDON · SINGAPORE · BEIJING · SHANGHAI · HONG KONG · TAIPEI · CHENNAI · TOKYO

*Published by*

World Scientific Publishing Co. Pte. Ltd.

5 Toh Tuck Link, Singapore 596224

*USA office:* 27 Warren Street, Suite 401-402, Hackensack, NJ 07601

*UK office:* 57 Shelton Street, Covent Garden, London WC2H 9HE

**British Library Cataloguing-in-Publication Data**
A catalogue record for this book is available from the British Library.

**Exploring Complexity — Vol. 8**
**BUYING TIME FOR CLIMATE ACTION**
**Exploring Ways around Stumbling Blocks**

Copyright © 2022 by World Scientific Publishing Co. Pte. Ltd.





ISBN 978-981-124-918-1 (hardcover)
ISBN 978-981-124-977-8 (paperback)
ISBN 978-981-124-919-8 (ebook for institutions)
ISBN 978-981-124-920-4 (ebook for individuals)

For any available supplementary material, please visit
https://www.worldscientific.com/worldscibooks/10.1142/12641#t=suppl

# About this book

> *"It is amazing how much you can accomplish if you do not care who gets the credit"*
>
> *Harry S. Truman*[1]

The book is the product of a process involving a small group of very committed people. Committed to steer humanity away from the largely self-induced and self-destructive course that it is on. The stakes are enormous: inconceivable suffering of billions of people and possibly an eradication of human civilization. To change that course requires urgent action. To determine which actions, a seasoned intuition is required for what can work in a given context and time frame. That, in turn, requires insights into the stumbling blocks that always emerge when actions aimed at change are planned.

To explore ways to avoid such stumbling blocks was what each participant in this small group of world class experts set out to do in the webinar: *"Removing barriers to buy time"*, that took place in February 2021. Annex 1 lists their names.

In this book, we have extracted their comments made during the presentations and discussions. These comments do not stand in isolation. They are expressions of the sharing of high-level knowledge and insights through three days of intense and in-depth discussions. Thus, the book is the result of a joint and intense effort. All participants deserve the credit for the thoughts, knowledge, interactions, wisdom and other contributions that shaped the contents of this book. The following persons should be mentioned and thanked in particular:

o *Carlo Jaeger*, who moderated the discussions on the last day of the webinar and who was instrumental in connecting some of the dots from the discussions in the previous days.

---

[1]  Truman, Harry (1884–1972) was the 33rd president of the US (from 1945–1953).

o *Jonathan Sim* and his team — Isaac Tan, Shayne Goh, Enoch Lim, Risa Pagdanganan, and Crystal Lee. Technically and technologically they "ran" the webinar and then provided the transcripts that contains the rough material in Section 2 of this book. Jonathan is also my co-editor.

***Buying Time for Climate Action***, *an exploration into ways around stumbling blocks* is about ways to deal with the mess that global climate change creates and the stumbling blocks that we encounter when trying to steer away from this mess.

The 2021 IPCC report[2], with its basic message that ***time is up, we need to move into the action mode as of now***, was a perfect pass for the production of the book. After all, this book is about getting into action and about the stumbling blocks that hinder plans to be turned into actions. The UN Climate Change Conference of Parties (COP26)[3], provided the perfect timing for its publication.

Jan W. Vasbinder
September 2021

---

[2]  Intergovernmental **P**anel on **C**limate **C**hange report: *Climate Change 2021, the Physical Science Base,* https://www.ipcc.ch/report/ar6/wg1/downloads/report/IPCC_AR6_WGI_SPM.pdf.

[3]  For more information, see https://ukcop26.org/.

# Contents

## Section 2: Exploring ways around stumbling blocks    **123**

# 1 Introduction

## Jan W. Vasbinder

In 1981, when I was a science attaché in Washington DC, a number of us were invited to attend a meeting of scientists who were worried about the $CO_2$ levels and their impact on climate. The meeting took place in a small room in the COSMOS Club on Massachusetts Avenue. It happened late in the evening and had a scent of conspiracy about it. Visionaries or prophets of doom who said $CO_2$ emissions must be curbed *or else*....

None of what we see happening now with the climate was visible 40 years ago. But these visionaries saw it coming even though they could not articulate very well what it was that they saw coming. Yet they knew it would be disastrous and they saw only one way to prevent it: limit the global $CO_2$ emissions.

The people, industry and government did not do that. We still do not, and climate change is happening. We cannot prevent it anymore. Yet we now know that it threatens to disrupt the vital balances that keep our societies going.

**Prophets of doom**
*Humanity has seen countless prophets of doom predicting its imminent demise. A classic example is Cassandra, who time and again predicted the destruction of Troje. Nobody paid attention to her, but she was proven right. More recently in 1798, Thomas Malthus theorized that, as the population grew exponentially while the food supply grew arithmetically, the demand for food would outpace its supply. Malthus predicted horrible famines and food wars and he was wrong.*

*Right now humanity faces a threat that has brought to the fore new prophets who cast their pall of doom. Climate is changing and in the worst thinkable scenarios, this may trigger an existential collapse of the social and technological infrastructure of our global community. Even though there are signals pointing that way, it does not have to happen. Malthus was wrong because he ignored technological progress. The new prophets of doom may be proven wrong because they ignore that, when the going gets tough, people at unexpected places step forward to meet the challenge.*

We know what we must do to prevent that from happening or at least to mitigate its consequences. And we need to do what is necessary *or else....*

**Early warnings**
*Already in 1974 climate modeller Syukuro Manabe presented model results indicating that a doubling of $CO_2$ levels would lead to a 3–4°C rise in global temperature.*

*Bert Bolin, who later became the first chair of the Intergovernmental Panel on Climate Change (IPCC) recollected the events in connection with an interview in 1991: "It was a remarkable result that to a large extent is still valid today".*
https://www.climate-policy-watcher.org/global-climate-2/bolin-bert-19252007.html

We have **known** that for more than 20 years, but we have been **warned** about the dangerous effects of rising $CO_2$ levels long before.[4]

In December 2016, Para Limes[5] organised a conference entitled, "Disrupted Balance-Societies at Risk"[6] that took place in Singapore. That conference dealt with the resilience of societies in the face of major disruptions. Twelve speakers, each uniquely qualified to address this issue, presented their views on the risks associated with major disruptions in systems that are essential to keep our societies going. The conference concluded that we could eliminate the existential threats to human society, if only we could "buy time" to delay the disruptions. The need to "buy" time indicates the urgency that was felt during the conference.

Yet, despite the urgency, we somehow do not seem to be able to transform that knowledge into effective actions. The reason is that we do not recognize or understand the complexity of the systems that we try to change. Nor do we see or appreciate the stumbling blocks preventing us from effecting the desired change.

---

[4]  "History Extra" provides an overview of what we knew about climate change and $CO_2$ at what time. For more information, see https://www.historyextra.com/period/modern/climate-change-warnings-history/.

[5]  Between 2011 and 2018 Para Limes was part of the Nanyang Technological University in Singapore, https://www.paralimes.org/.

[6]  For more information, see https://www.amazon.com/Disrupted-Balance-Society-Exploring-Complexity/dp/9813239212.

Human activities are constantly influenced by stumbling blocks that continue to evolve over time. Especially, it seems, these stumbling blocks are of our own making, and exacerbated further by our own actions. Attempts at dealing with such stumbling blocks through planned changes — whether in institutional arrangements, educational systems, political rituals, cultural patterns or other creations of the human mind — these plans never seem to unfold as planned, because *[...] in between dream and act there are hindering laws and practical issues and even melancholy that no one can explain.*[7]

In trying to get our act together, humans spend an inordinate amount of time struggling with such laws and practical issues or, more in general, with the stumbling blocks that stand in the way of actions, aimed at planned changes, to take effect.

> **Growing urgency**
> In the last five years the urgency has become a lot more evident. By 2019, the UN Climate and IPCC reports made clear that the global community must act within a generation to avoid an existential collapse of its social and technological infrastructure.[*] In August 2021, IPCC essentially reported that **time is up, we need to move into the action mode as of now**. The IPCC report stresses that immediate actions is necessary to prevent further warming of the atmosphere, ocean, cryospohere and biosphere.
> If no rapid, large-scale reductions in greenhouse gas emissions are effected immediately, limiting global warming to close to 1.5°C or even 2°C will be beyond our reach.
>
> [*] Essentially meaning that the infrastructure collapses without visible possibilities to recover.

---

[7] Willem Elsschot, 'Het Huwelijk', Verzameld Werk, Van Kampen en Zoon NV, 1957, p. 737. For an English translation of the poem, see https://puntkomma.wordpress.com/2011/04/13/the-marriage-by-willem-elsschot/.

> **Merchants of doubt**[**)
>
> *The scientific community has produced irrefutable evidence of the causal link between global warming and the related dangers and the increasing emission of greenhouse gasses. Yet a small, but potent subset of this community leads the world in a vehement denial of this relationship. These "merchants of doubt" run effective campaigns to mislead the public and deny well-established scientific knowledge. They form a formidable stumbling block on their own by blocking decisions necessary to curb the emission of greenhouse gasses.*
>
> [**) Oreskes, N. and E.M. Conway (2010) Merchants of Doubt. Bloomsburry Publishing, London, New York.

It is such stumbling blocks, that brought president Obama to sigh, when reflecting on his 2009 Cairo speech to the Muslim world: "How useful is it to describe the world as it should be when efforts to achieve that world are bound to fall short?"[8,9]

Addressing veterans on the 65th anniversary of D-day, Obama indicated why he believes change is possible: "Our history has always been the sum total of the choices made and the actions taken by each individual man and woman. It has always been up to us."[10]

It is up to us to effectively deal with stumbling blocks. The better we manage to avoid, bypass or handle those, the better we will succeed to shape a future that will not be dominated by gloom and doom, and instead be shaped by hope and bright perspectives. We have to move from analysis and aspiration to operation (from *dream to action*) and we have to do that now. That means we must develop action plans that can be effectively executed.

---

[8]   Obama, Barack (2020), *A Promised Land*, page 366, Viking Penguin Random House UK.

[9]   Obama's "sigh" continued: "Was Václav Havel correct in suggesting that by raising expectations, I was doomed to disappoint them? Was it possible that abstract principles and high-minded ideals were and always would be nothing more than a pretense, a palliative, a way to beat back despair, but no match for the more primal urges that really moved us, so that no matter what we said or did, history was sure to run along its predetermined course, an endless cycle of fear, hunger and conflict, dominance and weakness?"

[10]   *Ibid*, page 371.

To be effective, the actions[11] we plan must include ways to bypass stumbling blocks. To identify such ways was the purpose of the webinar, *"Removing barriers to buy time"*[12] that took place on 15–17 February 2021, with 23 experts[13] from many different fields. The ideas they discussed, and the knowledge they shared, form the core of this book.

It is our hope that this knowledge and ideas will help to deal with such stumbling blocks and contribute to action plans that can be executed effectively.

> **Climate Change**
> *Whatever the complexity of the systems that cause it, climate change is here to stay. Some of its consequences manifest themselves every day and everywhere, while other consequences may not have become obvious yet.*

---

[11] Action plans do not buy time, effective execution of the actions in the plans does.

[12] https://www.paralimes.org/events/past-events-video-recordings-and-slides/webinar-removing-barriers-to-buy-time-15-17-feb-2021/.

[13] Annex 1 lists the names of all participants.

# 2  A kaleidoscopic wealth of approaches

## Jan W. Vasbinder

This book is divided into two sections, separated by an interlude.

In the first seven chapters of Section 1, seven invited experts discuss a selection of the major problems facing humanity and point out what prevents us (and in some cases what possibilities they see) to enact any real change in the systems that cause those problems. The first of these presentations gives an overview of existential threats for humanity. The other six focus on five areas where the consequences of climate change already lead to (or are most likely to lead to) major disruptions (food, fisheries, water, infectious diseases, and forced migration), or where stumbling blocks prevent effective actions (the financial system).

Together, these presentations cover the consequences of our human actions (along with their relation to climate change) that may develop into existential problems if we do not change our ways and act accordingly. Implicitly, they point to the urgent need to adequately address these problems and find paths around the stumbling blocks (or outright eliminate them) because they impede our efforts to meet the urgency of acting in the face of the consequences of climate change.

Following these presentations there is an Interlude, consisting of three chapters. Chapter 10 offers an ancient, multicultural and evolutionary perspective and context when addressing change in general, the relationship between human-induced change and stumbling blocks, and for ways to deal with them. Chapter 11 provides an analysis of the nature of the stumbling blocks encountered when plans to deal with the consequences of climate change are to be executed. The chapter highlights the paradoxical tension that exists between our modern scientific methods that reduce the complexities of this world to meet our existential needs; and how the very same reduction of that complexity into a linear cause-and-effect relationship is itself a key contributor to many of our human-induced stumbling blocks. Chapter 11 also identifies the five main stumbling blocks that our panel of presenters alluded to. They are finance, talent, bureaucracy, vested interests, and political will and support (otherwise known as political potency). It explores that possibly the best way to deal with stumbling blocks is to find ways to avoid them, using multiple approaches to go around them. Chapter 12 presents a selection of highlights of the discussions in Section 2.

> **The nature of stumbling blocks**
> *Key are the phenomena that emerge from the interplay of the complexity of the systems that cause climate change and its consequences, the characteristics of the collectives that one has to deal with and the laws of evolution that set the context for what can be changed in a sustainable way and what cannot.*

Section 2 presents the enormous variety of ways in which stumbling blocks manifest themselves in practice and of approaches that are being explored to deal with them.

But maybe more relevantly, they also reflect that there is a need and a continuous struggle to find a balance between bottom-up actions and top-down coordination. During the webinar, it became crystal clear that there is not "one approach that fits all cases".

The seemingly unstructured sequence of chapters in this part reflects the almost **kaleidoscopic wealth** of approaches to deal with (or rather **go around**) the stumbling blocks that were discussed in the webinar. Each context highlights the need for an approach that is unique for that context. Rather than evoke the impression that there is a spectrum of approaches waiting to be followed, the participants demonstrated that effectively dealing with stumbing blocks requires on the ground expertise and networks, wisdom and a great deal of creativity and luck.

The 23 participants at the webinar represented such expertise, networks, wisdom and demonstrated a great deal of creativity. One can say with a great amount of confidence that their narratives, as presented in the chapters, present the state of the art.

One common thread, two conclusions and three recommendations emerge from these narratives.

The common thread:

- **It is naïve to believe that we can eliminate the disruptive power of stumbling blocks by attacking them directly.**

The two conclusions:

- **There is not one approach that fits all cases.**
- **There is an omnipresent need to find a balance between bottom-up actions and top-down coordination.**

The three recommendations:

- **Looking for ways *around* the stumbling blocks that emerge from the system must prevail over efforts to change the system generating those stumbling blocks.**
- **Fairness, perceived but also proven, is essential for all policy and bottom-up action to build trust enabling changing attitudes.**
- **Let thousand flowers bloom.**

# Section 1: Presentations

| | |
|---|---|
| **Martin Rees:** | Existential risks |
| **Tim Benton:** | Food systems and their sustainability: Buying Time |
| **Gert van Santen:** | Fisheries and climate change |
| **Daniel R. Brooks:** | What are emerging infections diseases? |
| **Hillary Brown:** | Climate-induced managed retreat, a multifaceted action plan |
| **Andrew Sheng:** | Finance as barrier to addressing systemic climate change |
| **Alexander J. B. Zehnder:** | Resilient water management |

# 3   Existential risks

## Martin Rees

*Martin Rees is Astronomer Royal and a Fellow (and former Master) of Trinity College, Cambridge. For ten years, Rees was Director of the Institute of Astronomy. He has received numerous international awards including the Balzan Prize, the Crafoord Prize and the Templeton Prize. He is a foreign associate of the US National Academy of Sciences, the Russian Academy of Sciences, the Pontifical Academy, and several other foreign academies. He served during 2005–2010 as President of the Royal Society and in 2005, he was appointed to the UK's House of Lords.*

***

Covid-19 has many lessons for us. Why were rich countries so unprepared? It is because politicians and the public downplay the long-term and the global issues. They ignore the maxim:

> "The unfamiliar is not the same as the improbable[14]."

Indeed, we are in denial about a whole raft of newly emergent threats to our interconnected world that could be devastating. Some, like climate change and environmental degradation, are caused by humanity's ever-heavier collective footprint. We know them well, but fail to prioritise countermeasures, because their worst impact stretches beyond the time-horizon of political and investment decisions. It is

---

[14]  Rees, Martin. *On the Future: Prospects for Humanity*. Princeton: Princeton University Press, 2018, 118.

like the proverbial boiling frog — contented in a warming tank until it is too late to save itself.

Another class of threats — global pandemics and massive cyber-attacks, for instance — are immediately destructive and could happen at any time. The worst of them could be so devastating that one occurrence is too many. And their probability and potential severity are increasing. Indeed, I fear we are, as I said in my book, "On the Future"[15], guaranteed a bumpy ride through this century. Covid-19 must be a wake-up call, reminding us — and our governments — that we are vulnerable.

Jan has chosen climate change as the focus of this meeting, but I would like to set it against a wider backdrop of future projections.

Humanity's collective footprint is getting heavier. There are twice as many of us as compared to the 1960s — about 7.8 billion, all more demanding of energy and resources. The number of births per year, worldwide, actually peaked a few years ago and is going down in most countries. Nonetheless, world population, is forecast to rise to around nine billion by 2050. That is partly because most people in the developing world are young. They have yet to have children, and they will live longer. And partly because the demographic transition has not happened in, for instance, sub-Saharan Africa.

Despite doom-laden forecasts by the Club of Rome[16] and Paul Ehrlich[17] 50 years ago, food production has kept pace with the rising population. Famines still occur, but they are due to conflict or mald-istribution, not overall scarcity. But to feed nine billion in 2050 will require further-improved agriculture: low-till, water-conserving, and Genetically Modified (GM) crops. And maybe dietary innovations: converting insects which are highly nutritious and rich in proteins into palatable food; cutting meat consumption; and using artificial meat.

---

[15]  *Ibid*, page 63.

[16]  The Club of Rome is an organisation with the goal of addressing an array of crises affecting humanity and the planet.

[17]  Ehrlich, Paul. *The Population Bomb*, New York: Ballantine Books, 1968.

To quote Gandhi: "Enough for everyone's need but not for everyone's greed."

Projections beyond 2050 are uncertain. If, for whatever reason, families in Africa remain large, then according to the UN, that continent's population could double again by 2100, to four billion, thereby raising the global population to 11 billion.

Optimists say that each extra mouth brings two hands and a brain. But it is the geopolitical stresses, the inequalities between countries and within countries that are most worrying. As compared to the fatalism of earlier generations, those in poor countries now know (via internet, etc.) what they are missing, and migration is easier. It is a portent for disaffection and instability. Wealthy nations, especially those in Europe, should urgently promote a mega-Marshall Plan[18] for Africa, and not just for altruistic reasons.

And another thing: if humanity's collective impact on land use and the climate pushes too hard, the resultant "ecological shock" could irreversibly impoverish our biosphere. We would be destroying the book of life before we have read it. Biodiversity is crucial to human well-being. But preserving the richness of our biosphere has value in its own right, over and above what it means to us humans. Great Harvard ecologist, E.O. Wilson said that mass extinction is the sin that future generations will least forgive us for.[19]

My Cambridge colleague, Partha Dasgupta, has recently completed a report on the economics of biodiversity[20] as the UK's input into the conference in China next November. Hopefully this will be as influential as Nick Stern's report on climate change back in 2006.[21]

---

[18] The Marshall Plan was proposed by George Marshall to provide aid to the western European countries that were badly affected after World War 2.

[19] Wilson, E.O. *The Creation: An Appeal to Save Life on Earth*. New York: W. W. Norton, 2006.

[20] HM Treasury, *The Economics of Biodiversity: The Dasgupta Review*, Partha Dasgupta, UK: HM Treasury, February 2, 2021.

[21] HM Treasury, *The Economics of Climate Change: The Stern Review*, Nicholas Stern, UK: HM Treasury, October 30, 2006.

So, the world's getting more crowded. There is also another firm prediction: it will get warmer. In contrast to population issues, climate change is certainly not under-discussed, but it is under-responded-to. And as we know, under "business as usual" scenarios, we cannot rule out that later in the century, really catastrophic warming, and tipping points, would trigger long-term trends like the melting of Greenland's ice cap.

Climate change is potentially a "global fever", which somewhat resembles a slow-motion version of Covid-19. For instance, both crises aggravate the level of inequality within and between nations. Those in the megacities of the developing world cannot isolate and medical care is minimal. They are subject to pandemics and they will also be victims of climate change. Likewise, the poorest people in the poorest countries will suffer the most from global warming and its effects on food production and water supplies.

But, to insert a bit of good cheer, there is a "win-win" roadmap to a low-carbon future. The prosperous nations of the north should accelerate research and development into all forms of low-carbon energy generation, and other technologies where parallel progress is crucial. Especially storage (batteries, compressed air, pumped storage, hydrogen, etc.) and smart grids. Perhaps a global set of interconnectors along the Belt and Road[22] and maybe worldwide to maximise the efficiency with which intermittent renewables can eventually be used.

The faster these "clean" technologies advance, the sooner will their prices fall. They become affordable to, for instance, India, where more generating capacity will be needed; where the health of the poor is jeopardised by smoky stoves burning wood or dung; and where there would otherwise be pressure to build coal-fired power stations.

This, in my view, affects priorities in countries like the UK with high-tech expertise. Here are some numbers: By 2050, if India had an economy like China today, it would produce 10 times the UK's present total emissions. Cutting our emission by the last 20 percent

---

[22] A long-term initiative by China that aims to accelerate economic development in countries along the historical Silk Road.

towards net zero is really hard, but we could more easily enable India to cut its emissions by an extra two percent, and thereby we will be doing more to cut global emissions. So we should help countries like India and sub-Saharan Africa leapfrog directly to clean energy using modern technology.

A word about new technology, generally. We should be evangelists for it, not Luddites — without it, the world cannot provide sustainable energy, nourishing food, and good health for an expanding and more demanding population. But many of us are anxious that technology is advancing so fast that we may not properly cope with it, and its misuse can give us a bumpy ride this century.

Advances in microbiology such as diagnostics, vaccines, and antibiotics offer prospects of containing natural pandemics. But the same research raises (and it is my number-one fear) the prospect of engineered pandemics. Regulation of biotech is needed. But I worry that whatever regulations are imposed, on prudential or ethical grounds, they cannot be enforced worldwide any more than the drug laws or the tax laws. Whatever can be done will be done by someone, somewhere.

And that is a nightmare.

We cannot rule out human-induced threats far worse than those on our current risk register. Indeed, we have zero grounds for confidence that society's fabric can survive the worst that future technologies, if misdirected, could bring. We need a trade-off in privacy, security, and freedom if we are to minimise those risks.

In other contexts, we will confront the ethical issues that arise in climate policy where it is controversial how much weight we should give to those who will live a century from now. It will also influence our attitude to global population trends.

I conclude by returning to nearer-term politics. Unsurprisingly, opinion polls show that younger people who expect to survive most of the century, are more engaged and anxious about long-term and global issues, and their activism gives ground for hope. And it is good news that there is no scientific impediment to achieving a sustainable world, where all could enjoy a life better than those in the West do today. We live under the shadow of new risks, but these can be minimised by a culture of "responsible innovation", especially in fields like biotech, advanced AI, geoengineering, and by reprioritising the thrust of the world's technological effort.

We can be technological optimists, but the intractable politics and sociology engender pessimism. There is a serious concern that environmental degradation, unchecked climate change, and unintended consequences of advanced technology could trigger serious, even catastrophic, setbacks to our society.

And, of course, most of the challenges are global. Coping with Covid-19 is plainly a global challenge. The threats of potential shortages of food, water, and natural resources and transitioning to low carbon energy cannot be solved by each nation separately. Nor can the regulation of potentially threatening innovations, especially those spearheaded by globe-spanning commercial conglomerates. Indeed, a key issue is the extent to which, in a "new world order", nations need to give up more sovereignty to new organizations along the lines of the International Atomic Energy Agency or World Health Organisation, etc.

What about the role of scientists? They have an obligation to promote beneficial applications of their work and to warn against the downsides. Universities should offer their staff's expertise, and their convening power to assess which scary scenarios — eco-threats or risks from misapplied technology — can be dismissed as science fiction and how best to avoid the serious ones. We have set up a centre in Cambridge to do just this.

Science advisers to governments, of whom I know a number, have limited influence except in emergencies. They must enhance their leverage through involvement with NGOs; blogging and journalism; and by enlisting famous or charismatic figures to aid their cause. Figures as diverse as the Pope, David Attenborough, Greta Thunberg, Bill Gates, and the media, can all amplify their voice. That is important because politicians will make wise long-term decisions only if they feel they will not lose votes by doing that.

I will end with a flashback right to the Middle Ages. For medieval people, the entire cosmology — from creation to apocalypse

— spanned only a few thousand years. They were bewildered and helpless in the face of floods and pestilences, and prone to irrational dread. Large parts of the Earth were *terra incognita*. But they built cathedrals — constructed with primitive technology by masons who knew they would not live to see them finished; vast and glorious buildings that still inspire us centuries later.

Our horizons in space and time are now vastly extended as are our resources and knowledge. Yet, we do not plan centuries ahead. At first sight, this seems a paradox, but there is a reason. The reason is medieval lives played out against a "backdrop" that changed little from one generation to the next: they were confident that they would have grandchildren who would appreciate the finished cathedral. But for us, unlike for them, the next century will be drastically different from the present. We cannot foresee it, so it is harder to plan for it. There is now a huge disjunction between the ever-shortening times-cales of social and technical change and the billion-year timespans of biology, geology, and cosmology.

"Spaceship Earth" is hurtling through the void. Its passengers are anxious and fractious. Their life-support system is vulnerable to disruptions and breakdowns. There is too little planning — too little horizon-scanning. We need to think globally, we need to think rationally. We need to think long-term, empowered by 21st century technology but guided by values that science alone cannot provide. And that is surely what motivates us all at this meeting. Thank you very much.

## Discussion

***Nik Gowing***: The science and alerts are clear and unambiguous, but how do we get both public and corporate to buy-in especially among the huge numbers who remain largely unconvinced. The existential reality and urgency are still not clear when there are struggles and fears facing us such as securing an income or surviving Covid-19, etc.

***Martin Rees***: I touched on those things towards the end of my talk, but let me just say that I think it is important to recognise the key role of the most influential opinion formers. For instance, the Pope's encyclical[23] in 2015 was very important in the lead up to the Paris Conference. And our secular pope, David Attenborough has had a big effect on alerting the public about pollution of the oceans and biodiversity. Bill Gates's book[24] is out this week and I think it is very sensible and does not underplay the big challenge.

There seems to be a gradual shift of opinion — at least in the rhetoric — from major companies, so I think we can just hope that this will get through to the public. And then of course, the politicians will follow them and be prepared to go along.

But I would like to also emphasise a point which I made perhaps too briefly. It is that there are win-win situations. It is hard to get people to accept a bare bones policy that involves hardship, but if we can have exciting new technology which can be spearheaded by countries that are experts in these areas, developed and deployed to provide cheaper clean energy to allow countries like India to leapfrog directly to clean energy, like they have leapfrogged directly to smartphones rather than landlines, then this is a win-win situation.

I think that we in the prosperous countries can do more. I like to say that we in the UK contribute about 1.5 percent of the world's emissions, but we contribute a lot more than that percentage in clever ideas. We should focus our research and development on these efforts to help the world and have a big incremental effect on India and Africa and collaborate with them to accelerate their transition to "Net Zero".

***Wilhelm Krull***: I feel that we also need to think about more interactive modes of communication with other parts of society, particularly involving civil society more strongly and maybe much earlier than we usually do. What do you think about that?

***Martin Rees***: I certainly agree with that because, as I say, politicians will only respond if they feel that the voters will support them. I

---

[23] Pope Francis. *Laudato Si'*, Vatican City: Vatican Press, 2015.

[24] Gates, Bill. *How to Avoid a Climate Disaster: The Solutions We Have and the Breakthroughs We Need*, London: Allen Lane, 2021.

recall a nice quote from Mr Jean-Claude Juncker of the EU saying, "Politicians know what's the right thing to do, but they don't know how to get re-elected when they've done it."[25] And if that's the attitude, then these things will not happen. So it is very important that the public should take these issues seriously.

Of course, locally they are going to be issues like whether we are going to need more nuclear power stations for baseload. That of course raises all kinds of issues to make nuclear power acceptable. So that is just one example where, clearly, you have got to have people debating things at a local level.

***Carlo Jaeger***: My question is very simple. Let us say on the dark side of things — such as the disasters we have to somehow avoid and recognize — what importance do you give to the problem of war, given the destabilisation of the global system of nations?

***Martin Rees***: Well, obviously, we know that there are now nine or ten nuclear powers, and there is a risk of a nuclear exchange. But in terms of the number of weapons used, it would not be so globally catastrophic as what might have happened during the Cold War. Where we were actually lucky, in retrospect, is to survive. But there could be a new stand-off between new superpowers later this century that ends less well. And there could also be local exchanges in the Middle East or India or Pakistan.

But my big worry is that we can have social breakdown caused by small numbers of people by obviously bio-engineered threats or cyberattacks. So, rather than massive nuclear attacks, it is possible to degrade the quality of society and destroy the fabric of society by the cumulative effect of cyberattack. If an attack cut out the electricity grid in a large region for a long period, anarchy would ensue.

---

[25] Buti, Marco, Alessandro Turrini, Paul van den Noord and Pietro Biroli, "Defying the 'Juncker curse': can reformist governments be re-elected?," *Empirica* 36, (2009): 66.

In fact, I quote in my book[26] from a 2012 US Defence Department report, where they say that a state level cyberattack on the electricity grid on the eastern seaboard of the US would merit a nuclear response. I think it is that serious and this is getting more serious because AI helps the hackers as well as the cyber security people, so I do worry about that.

Also, just to give another analogy, society is becoming vulnerable to any disaster. I mentioned in my book that if we had a pandemic — I was writing before Covid-19 then — if there is a one percent fatality rate, that could overwhelm hospitals and lead to a breakdown in social order. In contrast to the Black Death which killed half the population of some towns and the rest went on fatalistically.

What actually happened this year has been the death of 0.2 or 0.3 percent of the inhabitants of the UK, Belgium and the United States. We know how much pressure that has put on. The one percent threat is really, really serious. I worry about this cumulative effect that leads to instability and disorder.

One other thing: the nuclear threat is being aggravated because cyberattacks on the command-and-control system are a new class of threat to the nuclear arsenals of the different countries. So I am afraid there is a lot of bad news on that issue.

---

[26] Rees, Martin. *On the Future: Prospects for Humanity*. Princeton: Princeton University Press, 2018, 21.

# 4　Food systems and their sustainability: Buying Time

**Tim Benton**

*Tim Benton leads the Energy, Environment and Resources program at Chatham House. He joined Chatham House in 2016 as a distinguished visiting fellow, when he was also dean of strategic research initiatives at the University of Leeds. From 2011–2016 he was the "champion" of the UK's Global Food Security program which was a multi-agency partnership of the UK's public bodies (government departments, devolved governments and research councils) with an interest in the challenges around food.*

***

What I want to do is to build on some of Martin's comments and talk a little bit about the food system as an exemplar for the way we run our world. To diagnose the problem: our food system is the number one polluter on a global basis; a big driver of climate change from multiple dimensions; and is what underpins global ill health. More people are made ill and die as a result of poor diets and air pollution and so on. Our food system works for many — in providing our daily bread — but it comes with a whole lot of downsides.

Instead of focusing on diagnosis, I'm going to talk about why our food system is what it is, particularly the economic drivers and how we can perhaps address them.

Following the Second World War through to the Green Revolution,[27] there was always the underlying concern that our food system ought to provide more calories on a per capita basis, because, historically, food shortages were never far away. And by providing more calories, food prices would go down, which is surely an all-round public good. Based on data from the 1960s to the present day, what we see is exactly that: as we improve cereal yields, prices come down and per capita availability of calories goes up.

The notion that we should design our food system to increase the availability and reduce the price of food has led us to where we are now, but we have now reached the limits of that paradigm's rationality. Today, as we improve cereal yields, it drives down the price of food. As it drives down the price of food, it increases the economic rationality of wasting it, and it increases the availability of per capita food on a daily basis. This incentivises the consumption of excess calories, leading to a global epidemic of obesity. For every increment of yield today, the data shows a disproportionate increase in both food waste and obesity prevalence; a unit of incremental yield creates more than a unit increment in harm that comes from our food system.

We call this the "cheaper food paradigm[28]" in the sense that cheaper food is a kind of design criterion for our system. The cheaper food paradigm, is based on two main prongs: (1) intensifying agriculture and the research that goes into that, coupled with (2) the liberalisation of markets so that a country can export excess produce to parts of the world with less ability to produce.

This cheaper food paradigm is driving a whole load of negatives. And it interacts with other things in vicious circles. Each time we think the solution to food security or economic growth is to improve

---

[27] The Green Revolution refers to the period in the mid-1950s to 1960s where there was a great increase in food production due to investments in infrastructure and technology in a bid to overcome food shortages.

[28] Benton, Tim G., Carling Bieg, Helen Harwatt, Roshan Pudasaini and Laura Wellesley, "Food system impacts on biodiversity loss," Chatham House, February 3, 2021.

our yields and grow more food, it just makes many things worse. I will illustrate this by walking through the arguments.

We know the food system is responsible for about a third of the greenhouse gas emissions driving climate change. In the last decade, we have increasingly seen the impact of climate change on yields. As the yields go down, we say, "We need to intensify agriculture to make up for the shortfall," ratcheting up the arguments for intensification. As yields go down because of climate change, agriculture also needs to expand into new areas. That, coupled with the increased demand for ever-cheaper food, incentivises land-use change, impacting on biodiversity (and further driving emissions of greenhouse gases and climate change). Further, as we drive climate change, we are increasingly relying on land to claw back carbon dioxide. For example, from "Biomass, energy, carbon-capture and storage[29]" (BECCS). The more we intensify agriculture, the more we drive climate change, the more we need land for carbon-dioxide removal. Therefore, the more we need to grow our agricultural footprint.

As we drive the intensification of agriculture, we deplete soils, which drives the yields down, which makes us say, "Oh we've got to intensify agriculture further."

As we intensify agriculture further, we incentivise the specialisation of agriculture: we grow more and more, but fewer and fewer crops. 50 percent of the world's calories come from three crops, 80 percent of the world's calories come from eight crops. That means we are growing at a vast intensity and a vast scale relatively few things in monoculture: creating nature-free farming landscapes.

As we intensify agriculture, it reduces prices, so we waste more. That waste provides other challenges — such as plastic pollution — but the embodied emissions are very significant. The Food and

---

[29] A type of carbon removal process that converts organic material into storable and usable energy whilst capturing the carbon emissions during the conversion process.

Agriculture Organization (FAO) estimated in 2013[30] that "If food waste were a country, it would be the 3rd largest emitter."

As we concentrate on driving up the yields of the common commodity crops, grain prices decrease, allowing for the possibility to feed grain to livestock. This reduces meat prices, driving up demand for meat, thus driving climate change further and impacting on human health (both through direct effects of meat consumption, especially processed meat, and the indirect effects of consuming too many calories).

Moreover, as we concentrate everything on wheat, rice, maize, plus a few other crops, we have dietary convergence on a global basis where everybody is eating calorie-rich diets and not enough nutrition. On a global basis we grow far more calories than needed, but insufficient fruits, vegetables and legumes for a healthy diet.

And, finally, as we drive climate change, it reduces the nutrient content of food because plants grow in different ways, differentially allocating proteins and minerals to their seeds. Both contribute to malnourishment and reinforces the urge to "produce more food" and intensify the system further.

My point here is that what we have created is a complex system, inherent with the panoply of feedback loops and negative unintended consequences; but nonetheless based on the premise that producing more "food" is a "good thing", without recognising the potential for incentivising the production of calorie-rich food, in environmentally harmful ways, to excess. It is a Jevons' paradox,[31] writ large: the more we produce, the more prices fall, the cheaper it becomes. This incentivises new ways to consume, which reinforces demand. The food system is a complex system which is deeply, deeply dysfunctional; every time we think the answer is in growing more food, it just makes

---

[30]    Food and Agricultural Organization (FAO), *Food wastage footprint: Impacts on natural resources*, 2013, http://www.fao.org/3/i3347e/i3347e.pdf.

[31]    The paradox states that increasing efficiency in resource use will increase resource consumption instead of decreasing it as it generates more demand in the long term.

things worse. Furthermore, as a complex system, it is developed to be internally stable, so it is resilient to change.

And from a "solutions' perspective", as a complex system, there is no one thing that you could point at and say if we did that differently everything else would be improved.

## *What might drive change?*

The way I articulate it is that change is primarily driven through a triumvirate: (1) people — citizens or consumers — give social license to the market through accepting the way it works and buying the produce; (2) politicians make the rules of the market (and in turn, are licensed by the people), and (3) the market serves to deliver what is within the rules, what people license in ways to drive profits, and what is acceptable to shareholders and other investors. Change will not happen through any one of these mechanisms (people, politicians, or industry) alone, though we often frame arguments in terms of "The Solution is the responsibility of only one part of the system".

As Martin says, for change to happen, there is a need to make the politics less toxic: people want to be calling for change and not resisting it. If people's attitudes change, politics will change, and as politics change, markets will change. This becomes a potential for virtuous circles to arise as a three-way interaction.

The core dynamic is around people, politics, markets and their endogenous dynamics. But in addition, there are, of course, exogenous drivers and events that will put pressure on the system. These include geopolitical change — like some of the things that we have just been talking about with Martin; new technology, perhaps impacts of climate change (e.g. disruptions to trade networks, people movement), and so on.

I will go through in turn, people, politics, markets and external events.

**People** are increasingly recognising climate change. As Jim Hansen said in 1988 when he first testified in Congress, people will buy into climate change when they notice the weather is changing and start feeling the impacts of extreme weather. And people all over the world are recognising that climate change is not making their lives better with effects like floods, storms, and wildfires. US$148 billion was the cost of the California wildfires in 2018[32] and they had a similar fire season the following year.

We are getting to the stage where it is becoming patently obvious that the climate is changing, and not for the better. People are starting to say, "I do not want to see climate change happening anymore". So, changing attitudes are starting to bring in the long-term horizon — I am not thinking about my children and their children only, I am also thinking about my insurance bills over the next year or two.

People are also increasingly recognising the links between food systems, climate change and even emerging diseases. And I think Covid-19 has really started sensitising people to another level. People are also recognising that things like food have different associations with the climate. For example, if you compare 100 grams of protein produced from beef or 100 grams of protein produced from tofu, the footprint is very, very, very different. Beef emits 25 times more Green House Gases (GHG) on average and uses 75 times more land.[33] In other words, as people are starting to recognise climate change is happening and that food is one of its drivers, then people's attitudes to food, and the composition of people's diets may change.

Through Covid-19's lockdowns, the role of the environment for good mental health has been recognised as well. People are recognising that climate and diets, as well as biodiversity and diets are related. Increasingly, people are also recognising the relationship between ill health and diet. Certainly in the UK, over 70 percent of people under 65, who have died from Covid-19, have been obese. Eating too many calories and not enough nutrients makes you ill and more likely to be impacted by other diseases.

---

[32] "Researchers from Tsinghua University Describe Findings in Economics (Economic Footprint of California Wildfires In 2018)." Health & Medicine Week, January 22, 2021, 639.

[33] Poore, J. and Nemecek, T., "Reducing food's environmental impacts through producers and consumers," *Science* 360, (June 2018): 987–992.

In a first approximation, if we move towards a better, healthy diet, we would also have a diet that has a lower footprint and is more sustainable. In sum, the motivation from people to eat better, whether from a health, climate or biodiversity perspective, is increasing. This is likely to drive both the politics of food and the way the market responds. And, of course, one of the interesting things about complex systems is that there is scope for tipping points, where a linear driver creates a nonlinear change in behaviour that causes the system to re-organise into a different state. One of the things that we underestimate is that you can get social tipping points that work quite quickly. For example, a rule of thumb is that when about 20% of the people start doing things in a new way, it creates rapid positive feedback via peer pressure, converting much of the rest of the population. Hence, whilst dietary change may be seen, at the moment as a niche, it may not stay that way forever!

**Politics**: There is a need to change the drivers of politics. Over the last decade or so, whilst I have worked frequently with the UK and European governments, there is an increasing recognition that the costs of the food system are levied in different policy areas. In the UK, we have a food economy worth about 120 billion pounds. But when you add up the costs from poor diets to our health service, the cost of pollution to water and air quality, the cost of carbon dioxide, the cost of biodiversity loss, the cost of anti-microbial resistance silting up water courses — you get a similar order of magnitude of costs.[34] It is therefore starting to be recognised that the benefits of changing our production system and changing our diets, are of equal sorts of magnitude as what we see the economic value of the food sector today. When analyses are being done on a global level, you get similar sorts of results: our food system is costing us more than the money we make from it. This is, of course, incentivising us towards thinking about how to align policies across the whole of the government, in order to ensure "system positive" outcomes.

---

[34] Fitzpatrick, Ian, Richard Young, Robert Barbour, Megan Perry, Emma Rose and Aron Marshall. "The Hidden Cost of UK Food: Revised Edition 2019," Sustainable Food Trust, July 2019.

And, of course, there are multiple levers that policies can use to change attitudes to food, sustainability, or the environment, and Martin mentioned some of these that are indeed at the top. There are lots of things we can do such as: research and development, we can incentivise smarter land use, we can change markets, we can change our trade relationships, we can actually reduce waste and change behaviour, we can incentivise better food choices, we can even use planning law to say whether it is okay to put a fast-food restaurant outside a school, and so on. There are many, many, many ways that we can push our system in the right direction if we choose to.

Turning to **markets**, I think one of the interesting things is that, as consumers get more sophisticated and as citizens get more understanding, they want to see transparency in the market, and we increasingly have the digital tools to provide this transparency (such as blockchain). So rather than people going into a food store and buying food based on trust or on the hope that somebody is regulating it, there is the potential for "radical transparency". This might well drive business models, because focus groups currently suggest that there is a gap between how citizens think food is produced and how food actually is produced (particularly — "naturalness"). Transparency will also potentially drive investments, such as through the Taskforce for Climate-related Financial Disclosures[35] (TCFD), as well as some proposed ideas for a similar taskforce on nature-related disclosures[36] (TNFD), if these make it clear which parts of the market are driving negative externalities, and change the willingness of investors to invest. Such transparency may well dis-incentivise externalisation of environmental (or health) costs.

Finally, we turn to **exogenous events** that can drive the system. Of course, there are many. Even over the last year, we have had Biden's

---

[35]  TCFD was established to provide better market transparency regarding climate-related risks and exposure so that stakeholders can make more informed financial decisions.

[36]  TNFD aims to provide a framework for corporations to identify and report nature-related risks that have not been covered by the TCFD so that stakeholders can make financial decisions that would bring about nature-positive outcomes.

election, Covid-19, the impacts of climate change, etc. The World Economic Forum's Global Risk Report highlights a range of "trends and drivers" which drive societal and business risks. The 2018 report summarises the data in a diagram.[37] Around the outside of this diagram are mega-trends that determine, through their interaction, specific risks. These drivers include: the incidence of non-communicable diseases, the rise of nationalism, breakdown of multilateralism, geopolitical shifts between great powers, polarising society, rising inequality and so on.

This set of drivers are largely "destabilising". There is a whole lot of negative things that are happening and the longer they continue, the more likely the system gets destabilised.

The upside of the destabilisation of the future is that when the system starts to creak and break because something external has happened, that is also the time when radical system change can occur. It will be an opportunity to reconfigure the system to become one that better delivers human and planetary health.

> **Support global efforts to adjust national macro-policies, consumer attitudes and political support to better facilitate fast and fundamental adjustment of critical industries and sectors, such as agriculture, trade and urban development.**

In summary, there are multiple drivers of radical change that are with us: dietary health, climate change, biodiversity, and environmental impacts. The external environment is shocking the system, and when the shocks to the system occur, they loosen the lock-ins of the complex system's resilience. I think that change also requires the politicisation of issues, so that we signal, as citizens, that we want the system to change. So politicians hear that signal and engage with change, and not avoid it because they think they are going to lose votes in it. Finance and all the other things can help drive the politi-

---

[37] World Economic Forum, "The Global Risks Report 2018 13th Edition," January 17, 2018.

cisation of issues. And I think transparency is key and transparency is one of the very positives that is arising from a greater digitalisation of our supply chains. With the politicisation of these issues, telling people as it is and making things obvious, hopefully political and market change will come from that.

## Discussion

*Andrew Sheng*: Is this the classic policy paradox where politics try to resolve conflicts whereas economics based (on the rational man), come up with policies that politicians cannot implement without pain and losing their jobs?

*Tim Benton*: Yes, I guess in a way, but it is also our food system and climate change. In general, it is an example of radical market failure in the sense that the economic incentives are not there to allow rational man to make a rational decision. So we are not pricing in the externalities of the way our lives impinge upon our future lives. And if we did price in those externalities, the market would have a better chance of balancing consumption with sustainable demand. At the moment, we have driven demand, and incentivise demand, to the point where any country that chooses to externalise costs on the environment gains a competitive advantage and this drives a race to the bottom. So it is not economics per se that is the problem, it is the way that we are not pricing the environmental costs into the economics. That means that everything runs away from us.

> **Incorporate major externalities and costs of no/insufficient action in defining climate related policies and market regulations, and encourage a pro-active approach whenever possible.**

# 5  Fisheries and climate change

## Gert van Santen

*Gert van Santen was a fishery specialist at the World Bank. He initiated what ultimately became "Profish", the World Bank vehicle coordinating national and global technical assistance and funding to enhance the design and implementation of sustainable fisheries development. He supported the design of fishery sector development strategies; and he initiated and managed bank-financed marine fisheries and aquaculture projects in over 30 countries. After retirement, he explores how trade, financial aid and international politics influence fisheries management successes and*  *failures. He has been a founding Board member of the Ocean Learning Partnership in Newfoundland, which supported a specific ocean related curriculum at the province's high schools.*

***

This will be a view from the beach. Fisheries and climate change have similar origins, they are complex adaptive systems borne from human involvement in common property such as the oceans and the atmosphere. The oceans do not only provide fish, they absorb greenhouse gases and supply oxygen. These systems are linked. Excessive exploitation of fish and the release of greenhouse gases destabilize both of these as mankind borrows from the future. Reversal of current practices requires political action, driven by feedback at multiple levels.

Fishermen are mostly catching the bigger species in the oceans; almost 30% of the global catch of marine fish comprise 10 species.

About 10% of the biomass of individual fish populations are reported to remain at "underfished" levels. Excessive fishing efforts not only squeeze fish biomass and reduce broad benefits, they also change the ecosystem itself, sometimes substantially. Predicting such changes in complex marine systems, in combination with the impact of climate change, has proven particularly challenging.

Fisheries management — a misnomer as it aims to guide human behavior — has to deal with the collective action trap. For a long time, it has not been seen as a fundamentally political process to manage the human disruption of a complex system.

Most current fisheries management efforts aim to regulate fish catches or income benefits from a limited natural resource, applying scientific or bio-economic theories, optimizing catches or economic rent of catching individual species. Other equally important concerns, such as sector related employment, social safety nets, distributional fairness, the broader ecosystem impact or even inter-generational benefits, are mostly ignored. To better take care of those concerns requires an agreed long-term vision, the institutions and leadership to translate the vision into strategic political decisions, and models and processes that can better balance multiple interests. In too many countries, regulations created by top-down management are being implemented by weak siloed institutions that are separately handling policy, management of regulations, sector funding, enforcement, adjudication and science.

We have applied this one-size-fits-all approach since WW2. This long experience provides mostly painful lessons that may suggest what may or may not work when attempting to mitigate climate change.

Current fish exploitation, marketing practices and individual fish consumption levels differ widely. Globally, poor people still consume less than 25% of the well-off.[38] Mankind mostly perceives fish as a free natural "common property good" available to all. The many difficulties to introduce and maintain effective governance of fish resources

---

[38] Food and Agriculture Organization (FAO), *The State of World Fisheries and Aquaculture 2020: Sustainability in Action*, Rome: 2020.

and the oceans in general, reflect the difficulty to fundamentally alter that perception.

Wasteful use of fish is no exception. An extreme example is the annual ritual of killing bottlenose dolphins, where, in a European country, many spectators come to watch. About a quarter of the global marine fish production is being processed into fishmeal, an energy intensive product originally an essential protein component of poultry and pig feed. It is currently mostly used in the aquaculture industry, to feed fish. Losses during catching, processing, distribution and consumption of fish may well exceed 10% of the reported global production.

However, the history and practice of fisheries management does not only generate bad news. Huge quantities of protein grow in the oceans and this can satisfy human and animal demand for protein of marine origin to feed our swelling human population in the future. For example, fishermen are currently not catching mesopelagic fish species; your local fish shop does not yet stock lanternfish (*Myctophum punctatum*), or meso-fishburgers.[39] About 10 billion tons of mesopelagic species commute daily in the oceans, moving to the surface to feed and diving back to several hundred meters depths at daybreak to hide from predators in the dark.

Alternatives do exist for particularly energy intensive or environmentally damaging fishing methods like bottom-trawling. When ordering a fillet of sole at a fancy restaurant, few people realize that roughly six kilos of diesel fuel may be required to produce a kilo of sole fillets. Salmon culture releases about seven kilos of $CO_2$ for the production of a kilo of mature salmon steaks. Shrimp culture faces regular virus pandemics and requires fishmeal-based feeds. Shrimp pond construction has substantially reduced protective coastal

---

[39] ThoughtCo provides an explanation about the Mesopelagic zone and the fishes residing in it. The Mesopelagic zone is the region between 600 to 1000 meters below the surface of the ocean. Various species of animals reside in this zone such as fishes, shrimps, squids, eels, jellyfishes and zooplankton. See "Life in the Mesopelagic Zone of the Ocean," ThoughtCo, last modified July 3, 2019, https://www.thoughtco.com/mesopelagic-zone-4685646.

mangrove forests in tropical areas. Alternative, less intensive aqua-culture production methods have proven more robust. Replanting mangroves is a proven, natural way to reduce the impact of rising sea levels.

## *Lofoten Cod Fishery*

This fishery is an example of a successful, alternative approach to fisheries management. For several thousand years, people in Lofoten have been dependent on catching cod as their main source of protein. Cod spawn in the coastal waters during the winter and fishing governs most activities of the local community.

For political reasons, the community was forced to sell their fish to a monopoly of traders in Bergen, Norway during the Middle Ages till the beginning of the 20th century. When that stopped, they decided to keep the traditional management of their fish resources in their own hands.

On hindsight, this proved too optimistic. The introduction of industrial (trawl) fisheries and a major reduction around 1980 of cod stocks — exploited jointly by Norway and the former Soviet Union — strongly affected Lofoten fisheries. The small-scale fleet declined substantially and suffered economically through a barrage of ever more complicated multiple administrative measures to rebuild stocks between 1980–1995.

Nevertheless, the local community continues to aim for long-term goals to ensure resource stability and maintain at least part of their traditional society and culture, their social safety net, even technology and markets. Expanding production is no longer one of their goals. Because of the Lofoten's almost colonial history, political interference in resource management from Oslo is — after the administrative management upheaval — currently stabilized. Fishermen receive scientific support, but local experience remains important. The community has a say how much fish to catch, and how to process and market their fish. The drivers of this whole system have been the local community, history (to some extent), but also political support for its demand to be considered a special fishery, where management will be based on multiple considerations.

## Japan's fisheries sector co-management

Japan has also been highly dependent on fish as the main source of protein for many centuries. During the Edo era (17th–19th century), governance of fisheries was divided into three categories: local, regional, and oceanic. In 1901 and 1949 the country formalized the prevailing system, emulating German legislation.

At the local level, one cooperative covering a few villages manages all the coastal fisheries, aquaculture and most of the infrastructure. The system combines strong community leadership, social cohesion and fairness, and makes extensive use of traditional knowledge. Economic autonomy permits balanced levels of often complex resource exploitation and labor-intensive fish processing and marketing.

Regional and national fisheries are handled by different committees applying flexible systems for conflict resolution between various stakeholders and fisheries.

At the regional and the local level, the long-term understanding to maintain healthy fish stocks and a prudent fishing intensity is still widely followed, as is the acknowledgement that changing circumstances may ask for adjustments. When oceanic tuna fisheries increasingly failed to recruit new crew to work for up to 20 hours a day for six months at a time, the decision to largely withdraw Japanese participation in the fishery was jointly funded from national, local and provincial sources. It illustrated the clear understanding that adjusting to new situations requires efforts from everyone.

The Fisheries Agency has been a key driver of this unique management approach at the national level. It is an institution with particular standing in the country. In most countries, fisheries governance is the responsibility of often junior ministries, unlike in Japan. The rigid legal framework and commitment to long-term stability of regional and local communities have also been critical for the success of the governance system. Private and cooperative industries have equally supported the current system. Not all is well though, the average

age of Japanese fishermen is reportedly close to 70 years old; so a fundamental adjustment of the sector cannot be avoided.

### *The EU common fisheries policy*

Compared to Lofoten and Japan, the EU's common fisheries policy has a brief history. It started in 1970 and gradually replaced policies of individual EU countries. It also reflects Brussels' focus on political compromise, science, institutions, rules and finance. The Brexit negotiations provide a clear example of its general governance model.

Two positive observations. First, The EU realized that it had to provide a frame for the sector's long-term future. That frame has not been accepted in all places, but the very fact that it is an integral part of a strategy to adjust what used to be a very overcapitalized sector is positive. Second, from the very beginning, the EU provided substantial funding (currently 0.75% of the entire EU budget) to address the downsizing of the fishing fleet. Some of that money has been misused, but at present, funds are better targeted to support a more resilient fishing sector.

Now, two negative observations. First, the start-up and adjustment process of the EU's involvement in the sector has been slow at multiple levels, which proved a key weakness of the Brussels approach. This has caused factors such as uncertainty, the lack of trust and fairness to constrain more effective implementation of the EU policy.

Second, although the EU has made considerable efforts to broaden the analysis of the impact of its fisheries policies, it still relies on a process of quota allocation governed by management theories that, while more elaborate than in the past, still focus on single objectives. In addition, such allocation has become entangled in an annual EU wide process of seeking political compromise on multiple non-fishery issues.

The European Parliament has been very active in pushing the common fishery policy process along. NGOs have played a useful role, while national producer organizations in some countries have been particularly effective in accepting compromise while translating the process at the local level. Last but not least, the media have done quite a bit to raise social awareness on environmental and cultural issues. As can be expected with many different, often vocal stakeholders, some pursue mutually exclusive goals. The EU's common fishery policy process shows that a rule-based approach, persistence, money, short-term strategy adjustment and time can overcome an increasingly demoralizing status-quo.

## West Africa (2013–2020)

Now, a few observations about West Africa, which enjoyed considerable fish resources between Mauritania and Nigeria that historically have been caught by small-scale fishermen in their famous pirogues. Since the 1960s, growing desertification forced many small farmers to move to the coast; men ended up as crew on the pirogues and women processed fish. Currently about five million small-scale fishermen in the area catch fish for mostly local and regional markets.

At the same time, improvements in fisheries management and political developments forced particularly older industrial fishing vessels to move elsewhere. Many ended up engaging in reflagged fishing[40] in West African coastal waters where at present about 500 industrial vessels fish, mostly illicitly, for export. Fish populations in West Africa's coastal waters have been increasingly over-exploited as a result and West Africa has been importing about 1.5 million tons of frozen fish annually for decades. After decolonization, individual countries created small projects to mitigate the resource over-exploitation and local market supply problems. The piecemeal approach did not really work. The World Bank, along with other donors, initiated a regional program about 10 years ago (in which I participated), aimed at exploring how the common sector problems could be better addressed.

The program operated at many levels and its experience has been mixed. Recent reports suggest it has been quite successful in moving the industrial fleet out of coastal waters. In the past, one could see

---

[40] Changing the country flag on which the vessels originally sailed out with to another flag so they can fish at certain regions of the oceans before changing it back to the original flag when they return.

the lights of many industrial fishing vessels fishing close to the beach. Now, in most countries, the six-mile exclusive coastal zone which is reserved for the small-scale pirogues, is mostly being enforced.

In this particular case, satellite technology proved critical, as it provides fast internet coordination between various parties — external, regional, and local political leadership. Much still needs to be done to create a healthy fishing sector in the area, but recent developments are moving in the right direction.

Multiple "agents" have been active in the complex system of West African fisheries such as local small-scale fishermen, local and international media, some parliaments and politicians, NGOs, and the catalyst function of the World Bank. The experience suggests that the timing of implementing measures appears critical. Sometimes, waiting brings unexpected benefits, at other times you have to move fast.

### Lessons from the history of fisheries management that may help efforts to mitigate climate change

- seek acceptance of a long-term vision (including key adjustments and solutions) early on;
- create permanent multi-level pressure to encourage change;
- identify and support drivers that can move key stakeholders and the general public;
- accept multiple narratives of change, even paradoxes;
- adjust strategies when stuck;
- reduce uncertainty and transparently seek fairness to enable faster acceptance of the need to change.

### Discussion

*Wilhelm Krull*: Thanks for the Lofoten example. How were the interfaces between science and the local communities managed?

*Gert van Santen*: Yes, the interchange between science and the communities has been an endless little war. Scientists have had difficulties with fishermen for a very long time, and on the other hand, fishermen have even more problems with scientists for also a very long time. There are exceptions, thank goodness, and I think the whole situation is improving. But for a very long time, these guys did not really communicate. They might have talked to each other and have a coffee together on the boat, but really talking about what they were doing and what the differences were has been a very, very difficult process.

***Jan Staman***: (To Gert and Tim) is there a leading theory in agro-economics on change and transition considering the low-price paradigm?

***Tim Benton***: Just to give a brief answer, when I tried to get the argument published, a group of economists told me that it was heresy to think like that. My background is in thinking about nonlinear systems and complex system change and all the rest of that, whereas most economists tend to think in terms of linear systems. So actually thinking through the complexity of this area is slightly anathema to current economic thinking, but Dasgupta's great report[41] does give a potential framing for thinking through the consequences of internalizing externalities in more sophisticated ways than people normally do. So, no is the answer, I think.

***Gert van Santen***: As already mentioned, in both cases, the models that have been used by biologists or economists to make bio-economic models, are optimizing models focusing on a single objective. As I have said before, in a complex system that is totally ridiculous. The science that we have used or the models that we have used really need to be adjusted. And I agree with Tim; this linear thinking in economics is ridiculous. We have to get away from it, and I also agree with the Dasgupta report,[42] which I hope will get a very wide reading because I think it is very important.

---

[41]  HM Treasury, *The Economics of Biodiversity: The Dasgupta Review*, Partha Dasgupta, UK: HM Treasury, February 2, 2021.

[42]  *Ibid.*

# 6   What are emerging infections diseases?

## Daniel R. Brooks

*Daniel R. Brooks is an evolutionary biologist whose work ranges from field studies in tropical wildlands to foundational studies of evolutionary theory. He now focuses on integrating fundamental evolutionary principles into proactive and effective action plans for coping with the challenges of global climate change, with particular focus on emerging infectious diseases (EID). Dan is Professor Emeritus, University of Toronto, a Fellow of the Royal Society of Canada and Fellow of the Linnaean Society of London.*

***

I am going to talk about emerging infectious diseases. But first, I want to give a sense of what we actually mean by emerging infectious diseases. It is much more than just a few viruses affecting human beings and making headlines periodically. The whole scope of emerging infectious diseases are all the pathogens affecting us and every species upon which we depend for socio-economic reasons. This includes diseases we have never seen before but are seeing now, as well as diseases that we thought we had contained or eradicated that are coming back.

This number — $1,300,000,000,000 per year — represents the annual treatment costs and production losses due to what are called the high probability/low impact emerging infectious diseases. These are the ones that are ongoing all the time that do not make the headlines. These costs are greater than the Gross Domestic Product (GDP) of all but 15 countries,[43] and that is a value that was generated before the Covid-19 pandemic, of course. In addition to being costly in their own right, this *drip, drip, drip* of costs in the high probability/low impact emerging diseases is not only expensive, but it weakens our capacity to cope with the low probability/high impact emerging diseases like Covid-19.

There are a number of things that we have discovered that most people, especially people in the policy realm do not understand and we think they should understand in order to see how to cope with the emerging infectious disease crisis in a different way.

### 1. *Pathogens are not "Enemies"*

The first is that we have to stop thinking of pathogens as enemies. They are not enemies. They are natural components of the biosphere, and their effects on us are not a personal insult to us. This is simply part of how they make a living.

### 2. *Pathogens do not magically appear and disappear*

The second is that pathogens do not magically appear and disappear; they are present even when they are not producing disease or head-lines. The re-emergence of Ebola[44] is a perfect example. The World Health Organization announced that they had eradicated Ebola. Of course, that is not true. What they meant was that there were no active Ebola cases in human beings for a month, but Ebola as a pathogen had not disappeared from the planet and has not magically reappeared.

### 3. *Colonizing new hosts does not require new genetic capacities*

The third is that when a pathogen colonizes a new host, which is the essence of emerging diseases, this kind of phenomenon does not

---

[43] Brooks, Daniel, Eric Hoberg and Walter Boeger. *The Stockholm Paradigm: Climate Change and Emerging Disease.* Chicago: University of Chicago Press, 2019, 14.

[44] World Health Organization, "Ebola virus disease – Guinea," World Health Organization, February 17, 2021, https://www.who.int/csr/don/17-february-2021-ebola-gin/en/.

require new genetic capacities. In other words, pathogens do not have to evolve new capabilities before they colonize new hosts. It turns out that emerging infectious diseases are expressions of pre-existing capacities given new opportunities. There are always many hosts around the world that are susceptible to pathogens, but they are not being exposed to those pathogens on a regular basis, so we do not know that they are susceptible and potential hosts.

4. *Direct relationship between climate change and emerging infections diseases*

The fourth is that there is a direct relationship between climate change and emerging infectious diseases. For the last year-and-a-half, we have been hearing a constant drum of "we have global climate change" and "now we have Covid-19." But the reality is that these are not separate entities. The emerging infectious disease phenomenon is an aspect of global climate change in a very simple and fundamental way, and that is that climate change initiates movement amongst species. That gives opportunistic species like pathogens more opportunities to be opportunistic, i.e. what creates the crisis. This is what creates a very large risk space for emerging diseases.

5. *Human civilization created new opportunities*

The fifth is that human civilization has created additional new categories of opportunities for pathogens. Agriculture and domestication; sedentary living and urbanization; global trade and travel; migration and organized conflict have all increased the movement of pathogens and the movement of susceptible hosts among pathogens. This creates ever more opportunities for pathogens and diseases to emerge.

6. *We can be proactive*

Finally, we can in fact be proactive about coping with emerging diseases. This is mostly denied by current scientific paradigms on the assumption that we have to wait for a pathogen to accidentally evolve the capacity to get into a new host. And as that is something we cannot predict, the only option is to try to cope with it after the fact. However, if colonizing a new host is actually based on pre-existing capacities, we can anticipate how those pathogens might operate in new circumstances in conjunction with susceptible host they have never seen before.

## Proactivity should be foremost

We have to understand that proactive behavior with respect to the emerging disease crisis cannot simply be an add-on or a marginal thing to do. It must be foremost because when pathogens colonize new hosts using pre-existing capacities, once they arrive in the new host, they are then capable of evolving independently within the new host. At that point, new capacities may emerge that are not predictable. If we wait until that happens, unsustainably expensive crisis response is our only option.

## The DAMA (Document-Assess-Monitor-Act) protocol

We have proposed what is called the DAMA protocol[45] or what we call, "finding them before they find us." This is a combination of the grassroots to international level scientific and policy activities. The idea is that if we know what is coming, we can then buy time. While searching for long-term solutions, we can buy time because we can anticipate what the pathogens are going to do as they come towards us, our crops, and our livestock. We can act to mitigate their advance and we can act to mitigate their impacts as they approach us.

## A sense of urgency

We have to adopt a sense of urgency for this because that $1,300,000,000,000 per year is actually growing. It has not stopped just because Covid-19 is taking over the headlines. We actually have very little control over pathogen capacity, which is the evolved capabilities of pathogens to infect new hosts. But we can monitor, avoid, and minimize opportunities for encounter between pathogens and

---

45   Brooks, Daniel, Eric Hoberg and Walter Boeger. *The Stockholm Paradigm: Climate Change and Emerging Disease.* Chicago: University of Chicago Press, 2019, 225–251.

compatible hosts. And that is the essence of buying some time while we look for long-term solutions. Thank you.

## Discussion

***Andrew Sheng***: I think Lord Martin Rees made a very fundamental point: politicians act short-term, but the issues are all long-term. And Dan has made another important point. Buying time is all about timescales. Actually, if we are willing to spend on prevention, even just on the risks that we recognize, we can avoid very heavy future costs. But we are not doing this.

And why aren't we doing this? Because economists refuse to acknowledge this. They use this rational man's argument and price time the same without understanding that this huge time inconsistency involves costs. And since we do not know about the future costs, let us ignore it! All the politicians ignore the very expensive costs should accidents actually happen. They say, "I can't deal with it," "I don't have time to deal with this," "I have to deal with the urgent." It really is an issue that we have to address because this is the elephant in the room. We refuse to admit that the economist paradigm based upon Newtonian classical views[46] is completely outdated.

***Martin Rees***: I was also bold enough to mention concern about engineered pandemics, Daniel. Is that just fear-mongering? Or do you think that we do need to worry about monitoring them to minimize the risk of that happening by bad actors?

***Daniel Brooks***: Well, there is an evolutionary firewall against engineered biological weapons being very effective: Pathogens do not distinguish their makers from their targets. Pathogens are better at understanding that things like skin color and language are irrelevant when it comes to the biological unity of human beings. And another element that is talked about a lot with respect to things like biological weapons is the notion of biological terrorism. So, an example would

---

[46] Frydman, R. and M.D. Goldberg. *Beyond Mechanical Markets: Asset Price Swings, Risk, and the Role of the State*. UK: Princeton University Press, 2011.

be infecting somebody, having them get onto an airplane, walk through an airport, this sort of thing. This is the way that influenza and Covid-19 spread. Again, the major issue has been that in order to be certain that the punitive terrorist is infected and infectious, you have to wait until symptoms appear. At which point that person will likely be interdicted while trying to get onto an airline. So biological weapons in the evolutionary contexts are fairly clumsy tools and not easy to direct outwards.

I do understand that there will be governments that are going to be messing around with these ideas. But compared with the natural emergence of pathogens and our human technological impact on the planet intensifying those historical effects, I think that is much more of a problem than worrying about what we might be trying to do technologically. So, bioterrorism is sort of the *bête noire* of medication and vaccination. Most people think that vaccination is a great medical victory for humanity, but the reality is that the most effective 20-year period in human history for producing vaccines was 1880 to 1900. We have been falling behind in terms of cost to production and time to production on vaccines ever since then. And I think that there is an analogous situation with bioterrorism, bioweapons, and things like that.

There are probably people who are going to think about doing this, but I do not think it is going to be hugely cost effective for any government that is thinking about trying to implement something like this.

***Nik Gowing***: How much then have we been blindly living largely charmed lives and on borrowed time assuming the only option is optimistic?

***Daniel Brooks***: That is a really good point. We think that technology could always run ahead of the biology, but we have gotten to the point now where we understand that this is not true. We cannot expect that to work, and we must have something else. Which is essentially, a better understanding of the basic biology behind what is happening, this then allows us to mitigate the impact.

In other words, instead of trying what we call "medicate, vaccinate, and eradicate," the emphasis should be on coexisting and avoiding exposure. But that is very time consuming because it means that the scientific community is going to have to spend a lot of time working with local people and recruiting citizen scientists to help them document what is going on out there and explaining to them why they should be doing this. And these are activities that most academic scientists, most museum taxonomists do not want to be bothered with.

***Jan Staman***: So these zoonoses outbreaks are no longer acts of God. The states and governments can be blamed for new outbreaks in the future. Does that make them sensitive for more long-term policy?

***Daniel Brooks***: That is also a really critical point to make. The way that I have been expressing when I give talks to a more general audience, has been to say that we now have a good idea of what we could do to make things better. But it is going to be difficult to imagine how those would be implemented, fundamentally, which means that we have to begin cooperating with our neighbors. And especially, we have to be able to cooperate with people that we do not like very much. So, any country that has a traditional conflict relationship with a neighbor will somehow have to work together. They have to understand that sharing information and sharing expertise on mitigating the impact of emerging diseases is in everybody's best interest. And whatever conflicts they want to continue with, have to be with respect to factors other than emerging diseases.

Now, that is not saying that I have any great optimism that this will happen. But at least we know what we could do.

# 7 Climate-induced managed retreat, a multifaceted action plan

**Hillary Brown**

*Hillary Brown is a Fellow of the American Institute of Architects and Professor of Architecture at the City College, City University of New York, where she directs its interdisciplinary graduate program in urban sustainability. A former member of the National Academies' Board on Infrastructure and the Constructed Environment (BICE) and a fellow of the Post-Carbon Institute, Hillary is a senior research Fellow at the CUNY Institute of Urban Systems, where she directs its Regenerative Urban Systems Lab. Earlier, as an assistant commissioner, Hillary founded NYC's Office of Sustainable Design in 1996, producing both its High-Performance Building- (1999) and -High Performance Infrastructure Guidelines (2005). For her sustainability leadership around public works, Hillary was elected to the National Academy of Construction in 2019.*

***

This short overview introduces one of several potential responses to the not-so-distant threat of major population displacement precipitated by climate change. It will show how policies and programs for carefully managed retreat might advance *other* societal objectives simultaneously. To do that, I am going to try to connect the dots between two trends that give rise to *one action plan*.

The first trend, the ongoing failure to curtail greenhouse gas emissions points us towards unavoidable climate disruption. The second, the demographic phenomenon of rural to urban migration is not new. It is very much persistent and has caused marginalization and devitalization of rural areas worldwide, creating a sharp urban-rural economic divide. Many regions are being undermined by drought, deforestation and industrialization of agriculture. Together with extreme weather and other stressors, climate instability helps to empty out rural regions as many seek employment concentrated in urban centers.

This exodus contributes to a relentlessly urbanizing planet. It produces what might be termed as "hyper-urbanized" areas, densely crowded cities, and vulnerable informal settlements. I am going to suggest here that, by middle-to-late in the century, both trends combined may ultimately render many inland and coastal metropolitan areas unsustainable. Especially effected would be the low-density sprawl in deltas and river plains in industrial and emerging economies. This could set in motion unprecedented population displacement.

Some scientists suggest that even if we manage to curtail global warming by two degrees Celsius, unchecked climate destabilization may profoundly remap human settlement patterns. Some recent estimates claim that 800 million people from more than 570 coastal cities will be at levels of risk by 2050 from sea level rise.[47] Twenty to thirty percent of the world's land surface would also face desertification, which would cause the potential relocation of more than 140 million people across sub-Saharan Africa, South Asia, and Latin America in total. This could result in a displacement of an average of one person out of 45 worldwide as well.

Where will these displaced millions find accommodation? Internal climate migration certainly will be one inevitable reality, but it need not become a crisis if targeted action is taken now to better predict and prepare. The objective is to "buy time", and not wait until 2050 when levels of chaos from displaced multitudes may begin to become unmanageable.

---

[47] WBGU. *World in Transition: Climate Change as a Security Risk.* London: Earthscan, 2007.

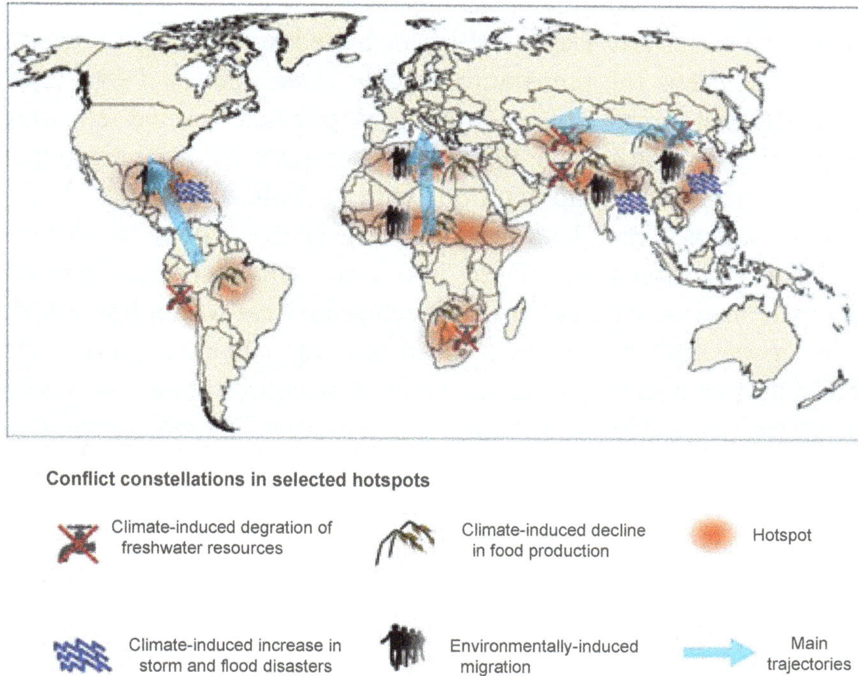



Fig. 1. *'World in transition: climate change as a security risk':*
*German Advisory Council on Global Change (WBGU), 2007.*

## Re-investment in the Hinterlands to accommodate retreating populations

Strategic managed retreat is one realistic action plan. Done correctly, it could address several problems at once. Managed retreat is conventionally considered as the last of three climate change adaptation options, coming after protection and accommodation. By definition, it is an effort to restructure development in climate-safer locales, replacing the at-risk settlements. This can mean deliberate resettlement of whole communities, which is what is examined here. Essentially it involves spatial planning scenarios that will likely take many forms to successfully blend the displaced with the receiving communities, and at the same time provide necessary services to support incoming populations.

However, the prospect of planned population resettlement at the scale just described leaves us in uncharted waters. One scenario explored here is the preparation of *climate-safe hinterland areas*, particularly, their small cities and towns as places to receive retreating populations. At face value, this is a bold proposition, but I argue that it is a dimensionally important one with multiple co-benefits.

For example, hinterland resettlement initiatives could prevent the influx of the climate-displaced into cities and megacities already overwhelmed or overburdened. If managed retreat within hinterlands were to be underpinned by societal reinvestment there, this could bring greater equity into these receiving regions, creating new jobs in service industries, bio-energy, and particularly food production. Arguably, such rural regeneration could ultimately result in the creation of more resilient and self-sustaining local economies. One could speculate that a more equitable and local way of life in small towns and small cities can help reduce both the inequities and the cultural dissonance found in today's urban-rural divide.

### *Revalorizing rural economies: Regeneration using circular frameworks*

With the slow onset effects of climate change, it could be said that there is sufficient time for preparation to occur in receiving areas. In the next few decades, investments could address rural decline and climate migration *both* by enabling hinterland regeneration to not only accommodate the climate-displaced but to also maintain the existing population in that locale. These designated climate-safe regions could be planned as a regional network of rejuvenated, compact small towns and cities. Moreover, if we take this opportunity to plan and re-develop these locales as "future-proofed" settlements, i.e. making these settlements immediately ecologically sensitive, decarbonized, and resilient, this strategy would realize both climate adaptation and greenhouse gas (GHG) mitigation simultaneously.

Such demographic renewal — purpose-built hinterland (re)developments — must by definition align with societal sustainability goals. This includes the regeneration of vital regional ecosystem services, and proactive rejuvenation of local economies to support arriving populations. Furthermore, a resettlement wave creates the opportunity for an intentional rebuilding of alternative "circular economies" in these townships. As an umbrella term used here, the circularity framework counteracts the ecological and socio-economic imbalances and the costs (externalities) of society's linear "take, make and waste" economy.[48] Instead, it is modelled on the tight workings of nature's metabolic flows. So a circular economy relies on services, products and material resources designed for *nonlinear or circular flows*, i.e. forming closed loops. What is currently perceived as waste is valorized by retrieving them as resources for reuse such as recovering waste energy, or repurposing used water. "Next generation" infrastructural systems are those optimized by colocation and interconnection such that waste output from one system provides free input to another, reducing systemic material, energy and transport costs and related GHG emissions.

A circular economy is also one that is predicated on forgoing fossil fuels in favor of local distributed and renewable energy. Closed-loop processes must be implemented across diverse production and consumption sectors, including regenerative approaches to local agriculture, fisheries and forestry. We have all of the readily available, off-the-shelf technical capabilities to do this. It is a question of how to optimize by reconfiguring the connections. In a hypothetical circular economy framework (Fig. 2), you can see how waste from one sector or activity becomes a beneficial input to another. The framework intentionally optimizes synergies, especially not just across the infra-structural sectors and commercial enterprises — as I show here — but also by incorporating regenerative approaches to local agriculture and forestry under the rubric of a more bio-based economy.

---

[48] The "take, make and waste" economy is a linear model of how goods are produced and used. As the name suggests, natural resources are taken from the environment and used for the production of goods which are then used for only a short period of time before being discarded. This contributes to the wastage of natural resources.

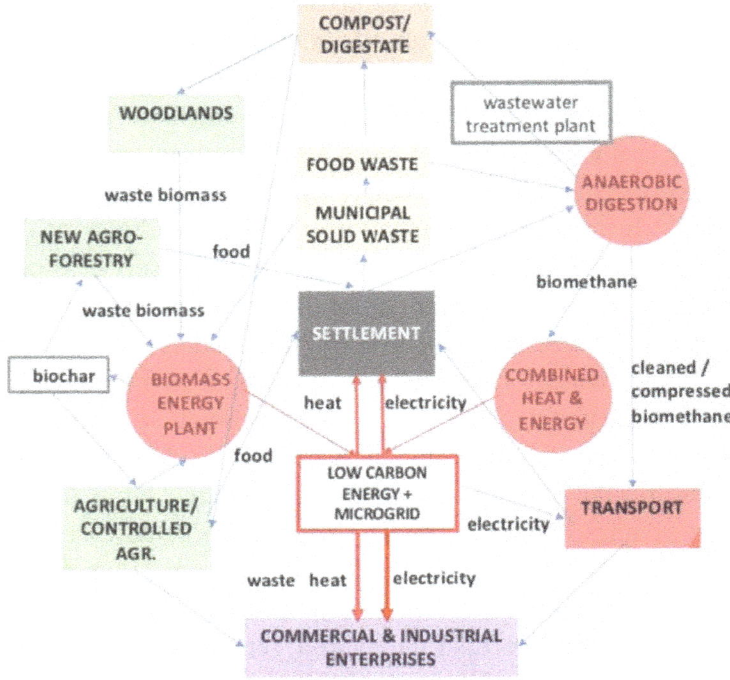



Fig. 2. *Hypothetical Rural Circular Economy (by the author).*

According to this economic paradigm, the economic and environmental regeneration of rural regions is not just an important end in itself. These efforts can also serve as a modest scale testing ground for the more challenging transition of large urban areas to the circular economy. Significantly, revitalization of these rural areas globally can reduce the consequences of the urban-rural economic divide.

## Multilevel enabling governance

The questions that logically follow from this proposition have much to do with how such a strategic retreat and resettlement program might be carried out. What kind of leadership is involved, and what are the barriers to be overcome? Starting with the notion of governance, we should anticipate the need for *multi-level modes of governance* to achieve a successful implementation.

First, there needs to be an overarching authority. Be it an international entity or a state body that can envision, incentivize and manage realignment of vulnerable populations. To do that, it has to be capable of creating a national discourse on the imperatives of planned relocation and identifying appropriate regions, while discouraging development in vulnerable regions. This has to be done at the state level to establish new institutional structures and domestic laws to first prepare receiving townships, and to operationalize resettlement subsequently.

To do this, the national government needs to provide detailed roadmaps and best practices. Currently, these do not exist, except in fragments and they have never been tested at this scale. National strategies, though, have to be complemented by a much more forward-thinking and nimbler regional or local government that will consider alternative modalities and scenarios for resettlement. They need to undertake the infrastructural realignments, and rebuild a bio-economy. Particularly, they also have to be capable of sharing power across what needs to be cooperative and inclusive networks of public, private and civic stakeholders.

## Challenges to managed retreat

Even as public awareness is attuned to impending climate chaos, there are multiple impediments — financial, political and socio-cultural — to undertaking strategic retreat at this scale. We know that internally managed retreat will certainly entail significant national costs that include buyout or compensation for loss private assets and fiscal incentives to relocate. However, this may be less costly than ongoing resiliency investments and the damage should those investments prove insufficient. Also, given our bias towards short-term political thinking, capacity for long-term planning for strategic retreat is lacking at all levels and across all sectors in both governmental and private sector management. Transcending this predisposition requires innovative leadership at both the state and local levels to communicate a compelling long-term vision that can mobilize resources around this transition. Lastly, we can anticipate that unless the receiving communities are integrally involved, they may resent and oppose an influx of outsider population.

> **Focus climate related strategies on the critical duality of the role of high level (national and global) policies, while providing maximum flexibility for the sharing of power and implementation of multiple, parallel actions at lower levels aimed at converting unsustainable production, trade and human settlement practices.**

## Conclusion

The transitions ahead of us in the coming decades are truly unprecedented. Unless planning and capacity-building begin soon, humanity is totally unprepared to deal with climate triggered displacement and resettlement at this scale, and the chaos it may bring. At the same time, there is an unacknowledged and unmet need to undertake redevelopment in the form of regenerating essential rural regions that have all too long been in decline. Combining these issues allows us to contemplate a multiprong solution serving both ends. I hope I have shown how anticipatory managed retreat could provide positive opportunities in comparison to reactive actions. At this juncture, we know that a post-pandemic world with high unemployment and the current discount rate hovering close to zero makes this an ideal time to explore this transition.

## Discussion

*Francis Vorhies*: Can the hinterland be both resettled and rewilded?

*Hillary Brown*: Yes, rural regeneration must simultaneously involve both. Promoting compact rural settlements helps conserve wild places which fosters biodiversity. Silviculture,[49] agroforestry, controlled

---

[49] Silviculture is the study and implementation of techniques to preserve the health and quality of woodlands and forests.

environmental agriculture, and permaculture[50] practices for local food and fiber production can all regenerate these areas, increase yields and rebuild soils — overall supporting biodiversity. And as these measures are relatively labor-intensive, they can create new employment opportunities.

***Gary Dirks****: What kind of capacity building efforts will be required to create a critical mass of skilled people and institutions capable of doing all of this?

***Hillary Brown****: There is a lack of capacity for the kind of horizontal thinking that is needed across sectors. In order to create a critical mass of strategic thinkers, we have to start with teaching complexity and systems-thinking, which is insufficiently stressed in the sciences or the humanities for that matter. I endeavor to teach some of these things to my students, but we really need to enable the next generation in a much more holistic manner while also giving them the leadership and problem-solving skills.

***Coleen Vogel****: To what extent does this work link to, if at all, to Kate Raworth's doughnut economy approach? Essentially, this is a very nationalistic model that she is using worldwide that shows the kind of outer threshold or limits by a physical world, and thresholds, the tipping points, and then the social, which is the floor — if you like — or the inner ring of the doughnut which are the social barriers or constraints. And then essentially what happens is that you have just got this nice juicy doughnut that we have got to keep and hopefully not transgress these boundaries. So, I think it does link into the circular economy approach, but it is getting huge traction. And I think even Amsterdam and their city council have adopted those.

---

[50] Permaculture is a design system developed to improve sustainability by creating an integrated agricultural ecosystem.

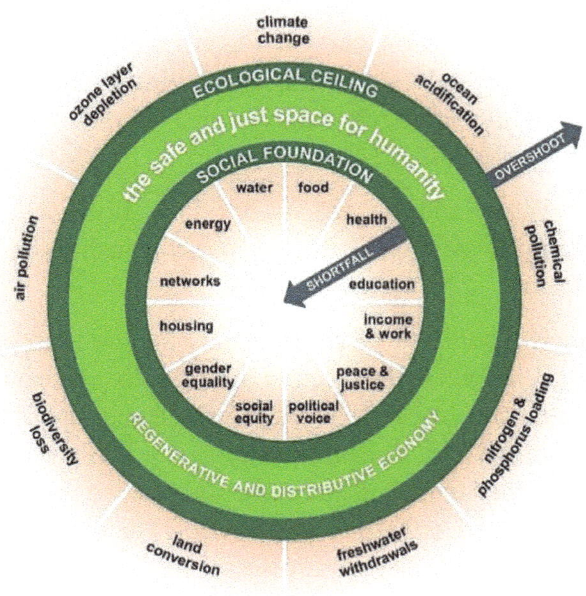


Fig. 3.  *Doughnut model: (Doughnut Economics).*

***Hillary Brown***: I absolutely think that these are both akin and complementary. You have to look at the socio-economic and political factors, which are highly intertwined, and which will permit us to *not* exceed those boundaries, but to regenerate from within. This doughnut model is another way of summarizing the integration of measures that enable us to stay within sustainable ecological limits, similar to say, a national eco-footprint. Similarly, the circular economy is an enabling concept to conserve and reuse resources to *avoid overshoot* and remain within that bounded band of a regenerative and distributive economy.

# 8 Finance as barrier to addressing systemic climate change

## Andrew Sheng

*Andrew Sheng is Distinguished Fellow of Asia Global Institute, The University of Hong Kong. He is Chief Adviser to China Banking Regulatory Commission, a Board Member of Khazanah Nasional Berhad and a member of the international advisory councils of China Investment Corporation, China Development Bank, China Securities Regulatory Commission, Securities and Exchange Board of India and Bank Indonesia Institute. He writes regularly on international finance and monetary economics, financial regulation and*

*global governance for Project Syndicate, AsiaNewsNet and leading economic magazines and newspapers in China and Asia.*

***

### Is finance a barrier to addressing systemic climate change?

On this topic, I am going to say *mea culpa*. I read economics at university, but I have come to realize that economics is the paradigm that is preventing me from looking at the issues as a whole.

Let me put it this way, until 2008, having worked all my life in central banking, financial regulation and finance, I thought I understood money. After 2008, central banking operations in quantitative easing and cyber-money[51] started changing all my concepts of how to look at money and finance. And in the last two years, I finally realized we are in deep trouble. This is because the economics paradigm is basically based on classical physics and a Newtonian mechanical perspective,[52] whereas the natural sciences have moved on to biology, quantum physics, neuroscience and the like. In concepts and mathematical tools, mainstream economics has remained largely in the old neoclassical world. As systems thinkers and ecologists Capra and Luisi have said, "As the twenty-first century unfolds, it is becoming more and more evident that the major problems of our time — energy, the environment, climate change, food security, financial security — cannot be understood in isolation. They are systemic problems, which means that they are all interconnected and interdependent."[53] We are all interrelated, we are all entangled, so how do we deal with this complexity of climate change, social inequality, finance, technology and geopolitics that are all interconnected and interactive with each other? The Chinese philosopher Zhuangzi (c. 286 BC) said, "Man and Nature are one". Of course, Einstein also said that "we cannot solve our problems with the same thinking we used when we created them."

So what are the barriers to changing our world and our thinking? Surprisingly, the biggest barrier is not just the physical but also the mental. What mental barriers are these? We have to go back to the philosopher Karl Popper, whose theory of knowledge categorized everything into Three Worlds[54] (Fig. 1).

The pre-scientific world thought everything was physical and had physical limits (World 1). For instance, the gold standard tied

---

[51] BIS. "V. Cryptocurrencies: Looking Beyond the Hype," BIS Annual Economic Report, June 17, 2018.

[52] Frydman, R. and M.D. Goldberg. *Beyond Mechanical Markets: Asset Price Swings, Risk, and the Role of the State*. UK: Princeton University Press. 2011.

[53] Capra, F. and P.L. Luisi. *The Systems View of Life: A Unifying Vision*. UK: Cambridge University Press, 2016, preface, xi.

[54] Popper, K. *Three Worlds*. The Tanner Lecture on Human Values, The University of Michigan, April 7, 1978.

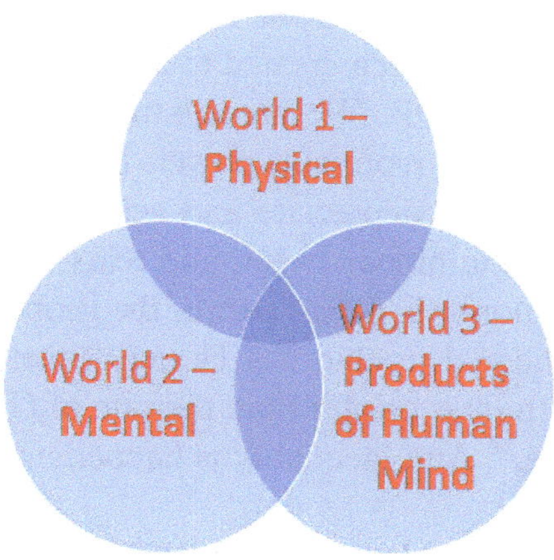


Fig. 1. *Money in Karl Popper's Three Worlds*
World 3 comes from the interaction between Worlds 1 & 2.

money to gold reserves. If there is no increase in gold, the supply of money cannot be increased. The return to the gold standard in the 1930s caused the Great Depression as prices fell and banks failed. To solve the shortage of liquidity, money and finance was broadened to include bank-created money in the form of bank lending. Karl Popper ascribed World 2 as the mental world because we have a physical one and a mental mirror of the physical world. For example, an asset could be physical, but a liability is a mental derivative of the asset.

But the mental world is a product of human imagination and human imagination is infinite. There are limits to the physical world, but our imagination is unlimited. The Popperian World 3 is the creation from all the products of the human mind: language, culture, institutions, etc. Suddenly, we discovered that if we can print money or borrow in excess; we can consume in excess. If we can consume in excess; we can consume the Earth or the planetary resources in excess, and that is the big problem that we have now. If everyone consumed like the average American or even a European, four Earths[55] are necessary.

---

[55] McDonald, Charlotte. "How many Earths do we need?", BBC, June 16, 2015, https://www.bbc.com/news/magazine-33133712.

What is finance? Finance (all the forms of gross money including financial derivatives) has been created to almost 10 times the amount of the annual world GDP. We are continuing to do this and the reason is because money represents power. If I can give you pieces of paper money which can depreciate and get physical resources from you or your services in blood, sweat, and tears, I am more powerful than you. And so money becomes part of the geopolitical domination toolbox. This is the real trouble. As Thomas Piketty said, if the rate of growth of wealth (r) is greater than the rate of GDP (g), inequality increases over time.[56] And inequality is happening big time.

## Global finance: Central banks becoming key driver of financial markets

How big is finance? Before 2008, global central banking assets were probably around US$12 trillion. Today, they have grown to US$40 trillion, larger than pension funds, insurance funds, and mutual funds. Just in 2020 alone, they created US$8 trillion.[57] What is US$8 trillion? US$8 trillion is roughly 10 percent of the world's GDP. Created just like that, with no cost and no constraints. They are doing it because we need to save the world from the pandemic, but what has happened? The richest one percent of the world's population has, according to Oxfam,[58] earned far more than the bottom 40 percent of the population, with poverty becoming worse after Covid-19. Society's inequality has become unsustainable and the trouble is that the one percent is controlling the media and controlling the paradigm and discourse. That is a serious social problem of capture, corruption and concentration.

---

[56]  Piketty, T. *Capital in the Twenty-First Century*. UK: Belknap Press, 2017.

[57]  Actual increase by four largest central banks (US Fed, European Central Bank, Bank of Japan and People's Bank of China, was $9 trillion in 2020. See Yardeni, Edward. Central Banks: Monthly Balance Sheets. May 19, 2021.

[58]  "World's Billionaires Have More Wealth Than 4.6 Billion People," Oxfam, January 20, 2020, https://www.oxfam.org/en/press-releases/worlds-billionaires-have-more-wealth-46-billion-people.

## Global Liquidity Trap: Too much short-term liquidity; insufficient long-term funds for Sustainable Development Goals (SDG)

So we have actually arrived at a Global Liquidity Trap. At the present moment, the liquidity of the world is so great that it is all trapped within the Wall Street, London, Europe, Hong Kong, Tokyo, and Singapore financial markets. With booming financial markets at a time of health and job crises, it is literally trapped there. And they are looking for investments, but at this strange time, the gap in long-term SDG investments is estimated by the UN to be US$2.5 trillion annually.[59] Compared with the US$8 trillion that the central banks can print overnight, there is hardly any long-term lending.[60] So you have a huge demand for infrastructure, spending on health, on urbanization, on food security, on disaster prevention but none of the liquidity is going to these urgently needed areas. It is all going and being lent at zero or negative rates to become funding speculation. The stock markets are higher than ever even though profits are hugely concentrated in some technological companies and the bottom half of society is really stressed.

Why is all of this happening? Central bankers have dropped interest rates (which was supposed to be the constraint on the unlimited creation of money) down to zero or negative. And the financial engineering solution to climate change is hyped to be about carbon trading and more green finance. But we know from the experience of the last ten years that a lot of the carbon trading and green finance benefit only the rich and the online crowd, which literally has very little connection with what is happening in the real world.

---

[59]   United Nations. "Citing $2.5 Trillion Annual Financing Gap during SDG Business Forum Event, Deputy Secretary-General Says Poverty Falling Too Slowly," United Nations, September 25, 2019 https://www.un.org/press/en/2019/dsgsm1340.doc.htm.

[60]   See Sheng, A. "Re-thinking Emerging Market Debt and Development Finance". In *4th Annual OECD – EMF Virtual Forum*, April 16, 2021, https://www.emergingmarketsforum.org/wp-content/uploads/2021/04/Rethinking-Development-Finance-and-Debt-by-Andrew-Sheng-1.pdf.

## **Conclusion: Money is not a constraint, but mindset is the barrier**

So the trouble is that the current financial system is unequal. It is very quantitative, it is predatory, and it is unsustainable. It is very sad to say this, but I was a central banker and I am ashamed of the way that we have completely ignored the moral issues by continuing to print money with no comment on where we are going to put that money. Finance is supposed to help the real economy and to allocate resources more efficiently. This is the strange part, to a large extent, for most of our careers and living memory, money was scarce. Today, money is not a constraint to buying time for climate change, but the mindset is. There is huge excess capacity due to the massive unemployment because of the economic recession. Central banks can engage in loose monetary policy because they think there is very little likelihood that, in the short-run, monetary creation will engender inflation.

So having removed that fear, why is it that most developing countries and advanced countries cannot invest in infrastructure?

The answer: politics.

For various reasons — one is very obvious — long-term infrastructure cannot be delivered because of collective action traps. But a collective action trap exists mainly because politics tries to reconcile conflicts and vested interests that engender polarization and policy paralysis. As we have recently seen in the 2020 American elections, 50 percent of the population is not listening to reason and there is anger out there. If you cannot persuade everybody, what happens to the policies? The policies over-promise and under-perform, resulting in huge distrust of the elites because the elites have failed the masses. You can call them deplorable or populist, but they have a legitimate voice which we are not giving them.

Thus, a paradigm change is necessary. If we do see that man and nature are one; science and arts are one; markets, state and civil society are one; we will need to switch to a paradigm that actually gets out of the current silo mentality that says it is always the other's problem.[61] As you have heard from presentation after presentation here, economics has failed to price or consider the externalities of our actions on others and the planet. And if we cannot quantify the externalities such as pollution or depletion of non-renewable natural resources, we do not price them in. Even if we price them in, the current paradigm thinks we can create a market to sort this out. And creating that market only benefits those who know how to benefit, and you do not get the change you want.

To a large extent, I have come to the conclusion that mainstream economics — I am not saying that all economists think like that (as Keynes said that policy makers are being enslaved by defunct economists) — have now dominated our paradigm so much that we cannot think out of the framework to enable us to get out of this collective action trap.

To sum up, money is not an issue, the question is: how do we change mindsets to free up money to effectively deal with climate change?

## Discussion

**Simon Zadek**: Andrew, in the past you have made the case that the global debt/equity balance is wrong, and that there must be greater focus on equity in real assets. Is this still part of how you see things in the wider context of your arguments?

**Andrew Sheng**: Absolutely, you know when I came back to Malaysia I got involved in Islamic finance and I finally understood Islamic finance. Islamic finance is about risk sharing[62] and debt finance is about risk transfer. Risk transfer means that you transfer all the burden of the risk to the borrower. If the borrower cannot pay, you send them to jail. But if the borrower owes you US$2 billion, you

---

61  Tett, G. *The Silo Effect: The Peril of Expertise and the Promise of Breaking Down Barriers*. US: Simon & Schuster, 2016.

62  Mirakhor, A. *Foundations of Risk-sharing Finance: An Islamic View*. In M.K. Lewis, M. Ariff and S. Mohamad (Eds.), Risk and Regulation of Islamic Banking, 107–128, UK: Edward Elgar Publishing Limited, 2014.

do not go to jail, the shareholder pays for the credit losses but the Central Bank prints more money to clean up your bad loans. So we literally have a very problematic gaming of finance.

What is equity? Equity means that if I put money in you and you lose money, I lose money. So it is actually extremely fair, but look at the global capital stock markets that I used to regulate. There are only 50,000 listed companies in the world. How many SMEs are there out in the world? Probably more than 200 million. Do they get access to become listed on the stock exchange? Almost zero — it is almost impossible for SMEs to get listed because you have to go to an investment banker and you have to hire very expensive accountants. The whole elite feeds out of this trough of creating money in which when they fail, the central bankers bail them out. This is the crime of the century. As a former central banker, I am ashamed to be part of this. This feeding is still going on.

> **Define and create flexible systems to expand financing climate related adjustment at global, national and particularly local levels, better reflecting the transparent and fair distribution of risks over time and reducing the window of the collective action trap.**

If we are able to shift more towards equity, then we realize that the return on equity could be negative because we are betting on the people. But if the masses get what they need — as we have seen through lending to women and lending to the poor — the returns are actually not bad at all. In fact, the social returns (on lending to the poor) are not bad at all, but the myth out there is that you must price that risk. You must price that risk into your lending rate, so that the lending rate for energy transformation for emerging markets, could be (as high as) 10 to 15 percent per annum, compared to zero or negative in the advanced markets. But this is a chicken-and-egg dilemma. If you do not put in the infrastructure, the poor keep on getting poorer, the rich countries will be affected, and we cannot solve a lot of the system-wide problems.

So there is a huge moral issue now, a moral hazard. We really need to think out of this current paradigm. We need an organic systemic perspective; a situation in which we all understand that we are part of the whole. That your problem is my problem; and we are all entangled. All this is fully understood in the natural sciences now, especially in the biological sciences such as the issue of emergence, the issue of chaos, the issue of order, and entanglement. But for economists, if it does not comply mathematically to my neoclassical model, you are not even counted. How many heterodox economic journals have tried to break through this barrier and cannot get through?

So as an economist, I think we are in the Reformation period. The cardinals of mainstream neoclassical thinking are so entrenched that if there is no mathematical model, the paper needs to be rejected. The result of this is that we have this very poorly understood (flawed) neoclassical model feeding into the policymakers' models which do not give them good solutions. In fact, the solutions that are not just bad, they are very damaging.

***Edward Kirumira***: Andrew, how does your argument pick on perhaps a non-mainstream Complexity economics idea that proposes a non-equilibrium view of the economy?

***Andrew Sheng***: Keynes tried in his 1936 General Theory[63] to argue against equilibrium. He argued that you will get into a liquidity trap where only government action can get you out of this. And then, of course, you know the neoclassical theorists like the Chicago school[64] came back with a vengeance in the 1980s and said, "No, the government is not a solution, it's the problem." Today, the market is the problem. However, the reality is that both of them are needed. In fact, on top of that, civil society is needed. And, as somebody said

---

[63]  Keynes, John M. *The General Theory of Employment, Interest And Money.* London: Macmillan and co., 1936.

[64]  The Chicago school of economics were advocates of free markets without government intervention. They believed that doing so would produce the best outcome for society.

earlier, it is not a two out of three trilemma, actually all three are needed to get to a solution.

We have been so transfixed about equilibrium that we (automatically) will revert back to equilibrium. This is intellectual nonsense. The neoclassical model describes an exceptional condition, not a universal one, especially in a system of dynamic change. Neoclassical economics has a pretence of (perfect) knowledge. Even Hayek argued against this — do not have a pretence of knowledge[65]; do not pretend that you know what you do not know. When he argued that the market had self-order, Hayek himself would have acknowledged that huge bureaucracies also have self-order.

The result is that we are in a new paradigm of non-equilibrium. And if we accept non-equilibrium — not just non-equilibrium, but multiple equilibriums, the policy thinking would be much, much better. I am on the Commission on Global Economic Transformation and a number of us went to see the Pope. And the Pope's Laudato Si'[66] suddenly woke me up because he said, you economists talk as if you can price everything, (but) you forget morality. And that is absolutely right. With interest rates at zero, everything is Islamic finance, there is no usury anymore at interest rates zero. Right? But all that money goes to the one percent which is hugely unequal. The result is the price of money at zero or negative is no longer an efficient allocator of resources. That part is extremely clear, but where does that put us? It puts us in a situation where it is going to wherever the politician says the money should go, irrespective of the huge waste that may be going on.

Let me give you an illustration. The United States has just given US$1.9 trillion to help the poor and I totally understand this. A very well-to-do couple, friends of mine, who are in the one percent of the population, have just received a US$2,800 government check which they do not need at all. So that is how bad the system has become. We have a complex situation of the ancient mariner's complaint

---

[65] Hayek, Friedrich. "The Pretence of Knowledge," Friedrich von Hayek – Prize Lecture, 1974.

[66] Pope Francis. *Laudato Si': On Care For Our Common Home*, May 24, 2015.

about liquidity or money: "water, water everywhere, nor any drop to drink".[67] That's happening to 80 percent of the population who need that liquidity but cannot get it.

***Jan Staman***: Could these oversized financial elites disempower and disintegrate the nation state?

***Andrew Sheng***: I do not think the nation state will go away, neither will cardinals change their religion, or even their stance. Having spent all their time becoming cardinals and I mean cardinals in every profession, they will not give up their fundamental beliefs. When that fundamental paradigm is challenged by the masses now, (it is) every man for himself. Sorry about this, but that is the reality. However, we can actually have collective action. I think to a large extent, the collective action trap is always solved, historically, by human society finding the answer. But it is not necessarily found by the elite. Ultimately, it is found by the masses, strangely enough.

---

[67] Coleridge, S.T. *The Rime of the Ancient Mariner*. US: American Book Company, 1834.

# 9   Resilient water management

## Alexander J. B. Zehnder

*Alexander J.B. Zehnder is a member of the Board of Trustees of Nanyang Technological University, Singapore. He is the founder and director of triple Z Ltd., Co-Founder of Nanrise Pte. Ltd., and Professor emeritus of ETH Zurich, Switzerland. He has been on the faculty of Stanford University, USA, Wageningen University, the Netherlands, and was director of Eawag, the Swiss Institute of Aquatic Sciences and former Scientific Director of Water Resources of Alberta Innovates — Energy and Environment Solutions in Edmonton, Canada. His work is focused on water policy, the nexus between water, energy and food security, water safety and infrastructures, as well as innovative solutions in industrial water treatment, particularly in the oil and gas industry. He also works on the development of capital market instruments for more sustainable production and use of natural resources. He has been a member of the team which developed the Dow Jones Sustainability Index and is one of the "founding fathers" of the "2000 Watt" Society, a strategic planning tool for more sustainable cities and countries.*

✳✳✳

## *Drivers of the water challenges*

As a society, humans are not only overusing and wasting our resources, but we are also largely managing their scarcities badly. Freshwater is a good example of an essential and limited resource. Responsible management can make the difference between sufficiency and scarcity. Local and regional hydrological cycles renew and redistribute water, typically in the form of precipitation (rain, snow). Aquifers, lakes, rivers, snowpacks, glaciers and ice sheets act as reservoirs. They smooth and buffer periodic insufficient or excessive precipitation. Desalination, re-using, or recycling of water increases the available volume of water, though at considerable energy and infrastructure costs.

The type and dimension of the water source defines the geographical extent for legislation and regulations. For basins, rivers and lakes, the dimensions are typically regional, national and multinational. In the case of virtual water (food supply) it can even be global.[68] The specific use of water determines the requirement for quantity and quality. Quality issues typically require action on a local to regional scale. They are rarely on a national scale, except for legislation and regulation.

Humans use water for drinking, hygiene, and household activities. With development and industrialization, industries and services like hospitals and restaurants, etc. are added. For all of these, water in its liquid form (also called blue water) is used. Green water refers to the water used for food production and is primarily in the form of soil moisture. It is the most substantial volume of water consumption (Table 1).

Since each individual has a **distinct need for water**, population size and economic development controls the overall water usage. Improved economic situation of individuals typically shifts the diet to more meat. Regional and national lack of water is primarily compensated by food import. This makes sense, because food represents the highest volume of the total water used. Water recycling can help all sectors, although the water for plant growth cannot be re-used or recycled because it is temporarily and locally lost through evapotranspiration. To be fair, it should be noted that the main water supply

---

[68] For in-depth information about virtual water and its role in providing food for water scarce nations, please refer to Yang, H. and A.J.B. Zehnder. "Virtual water: an unfolding concept in integrated water resources management." *Water Resources Research* 43: W12301 (December 2007): 1–10.

Table 1. *The water requirement for human survival and activities. The numbers are from different sources.*[69]

| Water use | $m^3$ per person and year | |
|---|---|---|
| Drinking | 1–3 | |
| Cooking, hygiene & sanitation | 11–18 | |
| Other household activities | 30–90 | |
| Services (hospitals, restaurants, etc.) | 50–140 | |
| Industrial activities | 50–140 | |
| Food production (agriculture) | 1,200–1,500 (mixed diet)[a] | 600–1,000 (vegan) |
| **Total** | **1,342–1,791 (mixed diet)** | **742–1,392 (vegan)** |

[a] Mixed diet with 20% meat. Both mixed diet and vegan are calculated to cover 2,500 kcal per day.

for agriculture is green water, which can only be used by plants and not be converted to blue water. However, the reverse process happens during irrigation (blue to green). Roughly 85 to 90 percent of the global food production is based on rain providing soil moisture. Freshwater also serves ecosystems and recreation purposes which will not be discussed further here.

Climate change, in particular **global warming** affects primarily the hydrological cycles by shifting rain patterns seasonally and geographically. This increases extreme events like floods and droughts.

---

[69] Numbers in this table are based primarily on Zehnder, A.J.B., H. Yang, and R. Schertenleib. "Water issues: the need for Haction at different levels." *Aquatic Sciences* 65 (March 2003): 1–20 and Liu, J., H. Yang and A.J.B. Zehnder. "Global consumptive water use for crop production: The importance of green water and virtual water." *Water Resources Research* 45, W05428 (May 2009): 1–15.

The combination of **demographic** development and **climate change** are the key drivers of the water challenges facing humanity now. The future will aggravate the situation because the human population is still growing, and the predicted temperature augmentation will increasingly influence hydrological cycles from local to global. *Since there is no life without water, managing water wisely and responsibly in a resilient way, is the only way to buy humanity time to adapt to the expected forthcoming global changes.*

## *Water management*

Resilient water management comprises three elements: (1) managing the resources in terms of quantity and quality, (2) managing the usage of water in terms of quantity and quality, and (3) successfully managing trust and emotions in an equal and inclusive way when changes in water management/policies are necessary. For any management decision to become accepted, this last point is absolutely decisive. People instinctively understand the role water plays in development and their own well-being.

As both the resources and the usage of water strongly affect each other, such dependences need to be taken into account when establishing solutions. Management matters are further complicated because worldwide, about 60 percent of all freshwater runs within cross-border basins and only an estimated 40 percent of those basins are governed by some sort of a basin agreement. For transboundary aquifers, agreements are mainly absent.

1. *Management of resources*
The main water resources which can and should be managed are rivers, lakes and aquifers. Precipitation can only be managed once it hits the surface and reaches a river, a lake or infiltrates into the ground. The overarching issue with **rivers** is the conflict between the upstream part and downstream part of the river. The points of conflicts are typically upstream activities that lead to chemical, biological and thermal pollution, or change of flow regime by water withdrawal or water works. Quantitative issues include damming, excessive water withdrawal, production of hydropower or channeling

to prevent flooding. Qualitative aspects besides pollution are the assurance of unobstructed hydro-ecological passage for fish and small aquatic organisms.

Usually when rivers flow through one jurisdiction (e.g. Mississippi) or when the partner countries are similarly developed with comparable political systems, finding solutions and reaching agreements is possible. Like in the case of the Rhine Convention[70] (originally between the five countries Switzerland, France, Germany, Luxemburg, Netherlands, and later joined by the EU as a whole) or the Canada-United States Transboundary Water Governance for the Great-Lakes and the St. Lawrence River region.[71] The Rhine Convention became the base of the EU Water Framework and has been used as a blueprint for the Danube[72] and other transboundary river systems with mixed successes.

When partners are far from equal (either economically or in military power) and when hegemonial powers are involved, fair agreements are less likely. There are many examples of failures that are of different degrees. Here are just a few: the Rio Grande (USA and Mexico); Jordan River (Jordan and Israel); Indus (Pakistan and India); Mekong (with China as the very strong upstream country); Euphrates and Tigris (with Turkey controlling the river flows). It is not necessary for the more powerful country to always control the source in order to prevent a good working convention. It may also be downstream like in the case of Egypt and the Nile.

---

[70] The convention was signed in 1950 by the original five countries to analyze the pollution levels of the Rhine and to implement protection measures to ensure a sustainable usage of the Rhine.

[71] There are several treaties and conventions as well as agreements under the Transboundary Water Governance initiative between the US and Canada. Some notable ones are the Boundary Waters Treaty, St. Lawrence Seaway Project and the Great Lakes Water Quality Agreement. These treaties and agreements aim to set guidelines to ensure that both countries do not overexploit or pollute these shared water boundaries.

[72] The Danube River Protection Convention was signed in 1994. The goal of the convention was to ensure that waters in the Danube basin are protected and used in sustainable and equitable ways.

**Lakes** are part of catchments and basins. In general, they are mainly exposed to pollution affecting natural resources. Particularly, the natural resources affected are water quality, fisheries, loss of biodiversity and habitat destruction. The shared management of the Great Lakes by the US and Canada is still the exemplar of how transboundary waterbodies can be used relatively sustainably by the partners. *The water richness, the similar cultural background, development stage and political systems of both countries aided the built-up of this show case.* In other parts of the world, the conditions are less favorable and thus hamper the development and implementation of strategies to protect the water resources of big water bodies. Once we learn how to manage water basins properly, even inter-basin water transfers over very long distances and different jurisdictions may become possible. This unlocks considerable resources like the Siberian rivers for Central Asia or the Great Lakes for the US Southwest.

**Groundwater** is the largest source of usable, fresh water in the world. In many parts of the world, especially where surface water supplies are not available, domestic, agricultural, and industrial water needs can only be met by using the water beneath the ground. Even in the most studied basins in the world, the hydrological link between ground and surface water is recognized but only understood at the reconnaissance level. While the effects of groundwater (over)use and pollution may be contained within national boundaries, legislation and enforcement are often lacking due to the "invisible" nature of the resource. There are still massive technical challenges in predicting spatial and temporal changes in groundwater systems with increased use.

### 2. *Management of usage*

Groundwater is the main source of drinking water because of its relatively high quality and large availability. The increasing pressure placed on resources by agricultural and industrial activities, as well as anthropogenic contamination (and also natural contamination although rarely, e.g. arsenic[73]), dries up the resource or makes it

---

[73] Some of the other natural contaminants of groundwater include aluminium, barium and nickel.

unsuitable for direct use. Technologies exist to bring water back to drinking water standards. They go from solar disinfection for single persons or households, to treatment plants serving millions of people. Water distribution goes from walking to the well, fetching it from the street vendor or from a supply store or truck delivery, to large distribution networks. Distribution networks usually supply more than just drinking water and water for household needs. They also deliver to stores, services like restaurants and hospitals, commercial entities and also for firefighting. These multiple tasks increase the infrastructure's complexity and cost.

The main difficulty in managing the usage of water is the almost absolute absence of source control by locals, with some exceptions in Europe and the US. Even if the local communities understand from a historical perspective what are the precursory measures needed for a sustainable supply, the higher political and administrative levels overrule them in the interest of the greater good (whatever that might be) and rarely make or supply the investments needed to replace or fix contaminated or depleted resources.

3. *Managing trust and emotions*

Imagine a spring with clean water, a little gurgling mountain stream with fish, a large river with children playing on its banks, a summer house on a blue lake, a tap from which clean and fresh water is flowing. All these images give a good and happy feeling. The people responsible for these positive emotions will be trusted in all their actions to safeguard the status quo.

Now imagine a dry riverbed, an empty reservoir, a formerly nice green meadow, houses and farms devastated by floods, harvest in a field destroyed by drought, etc. All these examples will not leave anyone emotionless. They evoke sad feelings. There will also be no trust for anybody who is responsible for these disasters. There could be many reasons for failure such as the lack of foresight or wrong political decisions, neglected infrastructures, absence of investments, or inefficient bureaucracy. One of the most common wrong political decisions is to renounce a catchment wide management.

Humanity has learned early on that treating and dealing with water in an equal and inclusive way is the key to trust, and consequently to a stable society. Respect for water, and proper and fair handling of water has become an inherent part of almost all religions around the world.

### Lack of trust leads to conflicts

Conflicts do arise within states if the river management is delegated by the central power to provincial level. In this way a watershed, though mostly in one nation, is not managed in its integrity. There are many examples here, some of the best known are: the Colorado River, where the water is already overallocated in the US, so little to no water arrives in Baja California, Mexico. The dispute about the Cauvery River in India sprang from the conflict between Tamil Nadu and upstream Karnataka which planned to use more water for irrigation. Iran went from a basin wide management by the central government to a provincial management. The result is that the river Zayandehrood is uncoordinatedly dammed and mismanaged in such a way that hardly any water reaches Isfahan.

These conflicts become sizeable obstacles and challenges if there is an economic gradient along the river, such as rural/urban or agriculture/natural ecosystems, etc. When rivers are shared by different countries, the conflicts are exponentiated and finding solutions become highly complex or close to impossible.

### Water management as the base for human civilizations

Human civilizations and societies have risen in places with a strong need for water management. Examples are Mesopotamia (Sumeria) around 4,000 BC, Egypt (Gerzeh about 3,500 BC), Indus Valley (Harappan about 3,300 BC), Peru (Norte Chico about 2,200 BC), Yellow River (Xia about 2,100 BC) and Mesoamerica (Olmec about 1,600 BC). These cultures developed and flourished when the water was managed in all its aspects. When water management troubles arose, such as the soil becoming salty as a result of inadequate irrigation (Mesopotamia), or droughts cutting people from their life base (e.g. Olmecs), these civilizations collapsed and disappeared.

We can learn from history that managing water quality and quantity successfully stabilizes societies and supports development. The November 2016 issue of *The Economist* states that water is scarce because it is badly managed.[74] The point made is that there is a physical scarcity in areas lacking water. But there are more places, in so-called economically scarce regions where a better management would clearly ease or even eliminate water scarcity. Comparing water scarce regions with governance indicators[75] show a striking correlation between less well governed countries and countries suffering water scarcity.

## Managing the risks

Demographic development is still in the positive range. Thus, the demand for water is increasing. Global warming is progressing. Even if countries follow their promises from the 2016 Paris Agreement on climate change mitigation, adaptation, and finance, the global climate will change far into the 21st century. As a consequence, the vectors of the main drivers of water scarcity are pointing towards a worsening of the situation. Buying time through adaptation to this global change is only possible by thoroughly improving the water management in all its facets. *The Economist*[76] identified 12 top concerns in terms of their severity and risk. Seven of them, including the most severe, are based on inadequate governance. Four are on global change and only one stems from the water resource. There is no magic to managing these risks, but why is it not done? There are (potential) barriers, or stumbling blocks. However, none is unsurmountable. They are all manmade.

---

[74] "The dry facts; Water." The Economist, November 5, 2016.

[75] Kaufmann D., A. Kraay and M. Mastruzzi. The Worldwide Governance Indicators: Methodology and Analytical Issues, 2010. The Worldwide Governance Indicators are available at: www.govindicators.org.

[76] Economist Intelligence Unit. *Water for all? A study of water utilities' preparedness to meet supply challenges to 2030*, 2012.

## *Potential barriers*

There is no one silver bullet or a solution which will fit all the regions and jurisdictions. Solutions must fit the specifics of the regional climate and local cultures and histories. The five main barriers are outlined in the following.

### 1. *Institutional resistance*

Water resources are often multi-community, multi-regional, or even multi-national. In addition, watershed and administrative borders may differ and responsibilities are distributed among various administrative units with little congruency. The phenomenon of "this is not my problem" and "they should be the ones to do it", together with the possible need of creating new institutions, entirely paralyzes bureaucratic initiatives.

Supplying drinking water, even in poor neighborhoods may become a cash cow if done right. If in the hands of the administration, politicians may use this potential for being elected by promising low tariffs to the electorate. The missing finances will then hamper the development of the infrastructure in the poorer neighborhoods. It is easier to subsidize the rich than to support the poor. To overcome institutional resistances, social, financial, ecological, and political interests must be aligned.

### 2. *Lack of talents*

The direct link between resource protection and economic development are rarely understood because of the absence of obvious short-term payback. As a consequence, there is little to no education for water resource engineers, except in countries with a history of long-term disaster. Groundwater systems are highly complex. Their accurate modelling and systems-prediction is a very demanding inter-disciplinary field. Only a few universities are fit to provide education in this area. And unfortunately, the small-scale local- and politically-dependent drinking water operations limit job markets and career prospects.

Exceptions are cities and a small number of water industries. Drinking water is a highly interdisciplinary field comprising social, economic, engineering, chemistry, physics and biological sciences. Concentrating this knowledge in very few individuals with limited career prospects in the mostly small-scale operation is hard to impossible to sustain.

**Expand education and research capacity to develop human capital in private and public sectors to redefine the prevailing economic and political narrative to implement climate related civic, public and private/corporate policies, regulations and actions.**

### 3. *Finances*

Water at the source is free and individuals can collect it at no cost. Costs may only be incurred with large quantity diversions. And because of the absence of cost for small volumes, there is no incentive for investments on long-term quality and quantity protection. Moreover, investments in the water sector typically have a long-term horizon, spanning as long as 80 years and the cost of infrastructures for drinking water production and distribution is very expensive. This is not really an incentive for action, and it makes for a "wait-and-see" attitude. This only changes to immediate action when there are disasters, like flooding and severe droughts. And unfortunately, the total cost of damages and losses, together with the necessary infrastructure investments for adaptation, often total up to ten times more than the cost of preventive measures.

In contrast to water from the source, everybody pays for the service to receive drinking water, either directly or via some sort of taxes. The exploitation of private ground water sources does have its cost, as well. Irrigation infrastructure is heavily subsidized and there is no willingness to pay for the water for ecosystem and rarely for ecosystem services. Infrastructures for drinking water production and distribution are extremely expensive. The initial costs for infrastructure often exceed the possibilities of communities who basically manage drinking water in most parts of the world. Lacking trust in governments for managing properly the finance over decades prevents pay-back and investments through tariffs. Privatization of drinking water most often fails because governments control the tariffs (because of opportunistic reasons) which do not cover full cost. Maintenance of infrastructure is the first to suffer.

4. *Collective action trap (political will and support)*
Without disasters, there is little to no public pressure for longer-term solutions for adaptation. A possible win-win situation is hardly seen because the time horizon for the positive effect is far beyond a political election cycle. So there is no immediate reward! This is the reason why only in very few countries is water high on the political agenda. An exception is perhaps hydro power. There are clear economic incentives, like power independence. AND dams tend to become national monuments which can be inaugurated with much pomp and provides high visibility to politicians. There is little to no short-term political return for a thorough planning effort and little glory for constructing sewers and underground piping systems. The mostly underground constructions are invisible and not suited for pompous inauguration events.

5. *Vested interest*
Free discharge into the rivers and lakes provide immediate cost savings. However, they get in the way of long-term quality protection. Similarly, there is the dilemma between upstream and downstream, often leading to the tragedy of commons. Protection of water rights might lead to severe conflicts. Therefore, private water markets, including those of the street sellers are easier to protect. Administrative boundaries between different units hinder sharing income and costs. In an immature democratic system and a weak administration, there is little to no incentive for investing into long-term goals for a healthy community in the future.

## Final thoughts

By managing water and its usage wisely, the existing freshwater on Earth can serve a developed global population of over 10 billion individuals even with the complication made by global warming.[77] If done wrongly, any water related activity may trigger strong emotions. Skepticism against new solutions, technologies and those who implement them will arise. Intellectual arguments will rarely change opinions. Only robust trust can prevent such fundamental opposition. Trust can be created by inclusiveness, full transparency, fair legislation, decoupling policy making and implementation, and eliminating ideological commitments. Trust will also remove the usual barriers preventing change.

## Discussion

*Hillary Brown*: Are we making progress with integrated water management and dealing with all facets of the cycle as one?

*Alexander Zehnder*: From the technological side, yes. But less so from the political side. We have excellent and very successful examples in all parts of the world. The most successful ones are those on a small scale, based on local knowledge and managed by the locals. Bali is such a great example and Oman is another. Local knowledge and long-term experience, as well as having common stakes in a well-functioning integrated water management, are keys to success. Grassroot democratic structures have developed over centuries to assure inclusiveness and trust. Since these grassroot democratic structures are not welcome in some jurisdictions, the integrated basin management is reluctantly tolerated but it is more often put under political pressure.

---

[77] The background of this statement stems from a number of publications among them Zehnder, A.J.B. "Wasserressourcen und Bevölkerungsentwicklung." Nova Acta Leopoldina NF 85, 323 (2002): 399–418 and Liu, Junguo, Christian Folberth, Hong Yang, Johan Röckström, Karim Abbaspour and Alexander J. B. Zehnder. "A global and spatially explicit assessment of climate change impacts on crop production and consumptive water use." PLoS One 8, no. 2 (2013).

On a bigger scale, there are two rather successful cases. One is the River Rhine. Over 40 years were needed to come to an agreement between five countries and a multinational organization (the EU). The main driver was the pollution. The Rhine flows through the most industrialized part of the world. The downstream cities use Rhine water to produce drinking water. The same downstream areas are also important clients of the products from the upstream areas (economic incentive). The political slogan was to bring back the Salmon to the Rhine. This asked for a drastic reduction of the pollution and some ecological restorations along the river. The other case is the Saint Lawrence River. Getting an agreement between the US and Canada to manage the river in an integrated way took an even longer time compared to the Rhine.

***Andrew Sheng***: What is the conclusion regarding the privatization of water delivery so far?

***Alexander Zehnder***: The crux with privatization is that in many cases it is only a pseudo act. Let me explain what I mean by this. Privatization of drinking water has been decided based on opportunistic political reasons. Governments have promised private companies that they are free to set the tariffs in such a way as to cover production, investment and maintenance costs for the infrastructure. Because investment and maintenance have been subsidized by governments previously, they did not show up in the tariffs. The discontinuation of subsidies forced the private sector to drastically increase the tariffs, which led to substantial public protest. Governments were forced by the public to control and lower the tariffs but they did not compensate the private companies for the loss in income. Thus, to optimize the business, the companies cut back on infrastructure maintenance which was obviously detrimental in the long run. Almost classic examples of this are Buenos Aires and Berlin. The British Railway system had very similar problems after privatization. However, after severe accidents, the government started to subsidize infrastructure investments and maintenance again. But they left the operations and their costs in private hands, which is a good model for public-private partnership.

This model is now applied in many countries where specific public services have been privatized.

Privatization does not need to fail, if done right. Many cities in France and the UK are good examples for a successful privatization of the drinking and wastewater sector. Due to some drastic failures, privatization of water supply and water management has a very negative image. Bad management of the transition from public to private created severe negative emotions and led to a total loss of trust.

# Interlude

**Heraclites, Lao Tzu, change, evolution and stumbling blocks**

**The nature of stumbling blocks**

- **Reduced reality and science**
- **Science and complexity**
- **Complexity and stumbling blocks**
- **Five stumbling blocks**
- **Stumbling blocks and collectives**

**Selected discussion highlights**

# 10  Heraclites, Lao Tzu, change, evolution and stumbling blocks

## Jan W. Vasbinder



Lao Tzu and Albert Einstein lived more than 2500 years apart, yet both had similar observations and thoughts. In fact, much of what modern men think and observe, has been observed and thought about by ancient individuals.

One example is Heraclites, the Greek philosopher of Ephesus (ca. 535–475 BCE) who is credited with the thoughts that "Everything changes all the time" and "No man ever steps in the same river twice".[78]

Heraclites also said: "The one is made up of all things, and all things issue from the one".[79]

Another example is Lao Tzu (ca. 500 BCE), the Chinese philosopher who is credited with the text of the Tao Te Ching that starts with words that strongly resonate with Heraclites:

> *The Tao that can be told is not the eternal Tao*
> *The name that can be named is not the eternal name*
> *The nameless is the beginning of heaven and earth*
> *The named is the mother of ten thousand things*[80]

---

[78] πάντα ῥεῖ, ουδεν μενει or "Everything Flows, Nothing Remains". And δὶς ἐς τὸν αὐτὸν ποταμὸν οὐκ ἂν ἐμβαίης or "You could not step twice into the same river." Both lines quoted by Plato in *Cratylus*, 402a.

[79] Quoted by Aristotle in *On the World* 5 396b20

[80] Other translations speak of a *Myriad of things*, essentially meaning: *Everything under heaven and earth.*

**Tao Te Ching**

*Tao is often translated as "the Way" or "the Path". In the foreword to the 2012 version of the 1972 translation by Gia-Fu Feng and Jane English, Tionette Lippe gives the meaning "Natural Law" to Tao and "Strength" to Te. Ching, meaning "Book", Tao Te Ching then translates into: "The book of the natural way and its strength".*

*In his introduction to the same translation, Jacob Needleman states that Te refers to "nothing less than the quality of human action that allows the central, creative power of the universe to manifest through it". Then Needleman points to what may be interpreted as a stumbling block: "The ego, our ordinary "initiator of action" is an ephemeral construction, which is formed by factors operating far beneath the level of the source and which, in the enlightened state of awareness, represents a kind of blockage or impediment to the interplay of fundamental ordinary cosmic forces".*

Although nowhere explicitly, both Heraclites and Lao Tzu can be read as pointing to the oneness of the universe, to continuous change and to evolution. Also, according to Lao Tzu, the key to harmonious change is "Wu Wei" roughly meaning "Effortless action" or "Acting by doing nothing" or "Doing nothing but everything is accomplished", all aspects of the Darwinian theory of evolution.[81, 82]

Intuitively, doing nothing does not seem to be the way to deal with stumbling blocks, yet it provides a promising perspective on ways to think about dealing with them.

This is the problem we need to solve: Climate change threatens our social and technological infrastructure. To counter those threats requires time to plan and execute actions. Stumbling blocks prevent actions to become effective and make us lose time.

---

[81]  In Medium (April 3, 2020) Kenneth Leong argues that the concept of *Wu Wei* in Taoist philosophy is consistent with evolution theory, the idea being that living organisms evolve from the very simple ones to highly complex ones, simply through natural selection. The organisms adapt themselves to the environment they are living in, very slowly and blindly. The life forms that do not adapt well to the environment simply die out. For more information, see https://kleong54.medium.com/when-laozi-meets-darwin-5e854b8ff8ad.

[82]  The key to Darwinian evolution then and now is a theory of *coping with change by changing*. Paraphrased from Agosta and Brooks (2021).

**Darwin, biological evolution and stumbling blocks**
*For more than a century, evolution was perceived as a process driven by similar needs for resources necessary for survival among organisms and species, producing conflicts of interest when those resources were in short supply. Inevitably, some would flourish, and others would wither away. This made a virtue of embracing conflict directly, winners replacing losers, and the winners being deemed "the fittest." This was an unfortunate modification of Darwin's original theory, which emphasized that evolution was a process of selective diversification, not of replacement. There must be a way for the inevitable conflicts of interest to be resolved in a manner that allows more than one participant to survive. Animals and plants that find themselves outside the arena of greatest conflict are no longer in contention to be "the fittest." But being merely "fit" but alive has advantages — such beings have landed in a space where there are degrees of freedom to explore, leading to the diversification that characterizes the biology of our world. This renewed appreciation for Darwin's focus suggests that human societies have more chance to survive the existential threats facing them by finding ways around the stumbling blocks than by entering into direct conflict with them.*

Stumbling blocks manifest themselves when humans want to make change happen, and especially when we want to make changes happen faster than the natural evolution of the systems that are affected by those changes.

In other words: We can plan actions to change an essential system but, unless we include pathways in the action-plan to deal with the stumbling blocks we will encounter, the actions in the plan will not lead to the changes we seek.

What can we, humans, do about this? How can we avoid wasting time and energy?

Our answer: Once you recognize a stumbling block, **look for a natural way to go around it**. Or, to put it in a different way: Take the lessons of Darwin and Lao Tzu to heart.

**Lessons from Darwinism**
*Darwinism teaches us that: (1) the only sense of progress is persistence in time (survival); (2) the great majority of the traits species use for persistence and coping with the conditions of life are specific and conservative elements of inheritance; (3) if you persist long enough, you may come up with better solutions for coping with the conditions of life, but they will always be contingent and temporary, because the conditions always change; (4) there will always be unanticipated consequences no matter how good a solution appears to be; (5) there are severe penalties for overgrowth; these can be deferred but never escaped; (6) deferred penalties for overgrowth are like compound interest, the longer they are deferred, the worse the penalties; and (7) evolution conserves nothing.*

One such lesson is: Evolution is relentless.[83] It conserves nothing. Living systems that do not adapt to changing conditions, will go extinct.

---

[83] An important source for the thoughts in this section is Salvatore J. Agosta and Daniel R. Brooks (2020), *The Major Metaphors of Evolution, Darwinism Then and Now*.

# 11  The nature of stumbling blocks

**Jan W. Vasbinder**

The systems that keep our societies going are at risk of being dramatically disrupted. Climate change is seen as a common cause for these disruptions, but the underlying causes are the growth of the world population, the increasing global connectivity, the unexpected consequences of collective human consumption and behavior, and especially the interactions and synergies among these factors. The knowledge to avoid an existential collapse of our social and technological infrastructure is basically available; there is an urgent need to mobilize and apply it. However, implementating the necessary measures is enormously difficult, because a number of stumbling blocks prevent appropriate actions.

We see the problems unfolding but cannot get our act(ion) together to address them timely and in an adequate way (a working combination of money, talents, organizational flexibility, political pressure, tailored compensation or targeted laws and regulations). Stumbling blocks or *hindering laws and practical issues and even melancholy that no one can explain* stand between the intentions and effective actions.

Stumbling blocks are the cause when actions that seem adequate and reasonable, in view of the problem that needs to be addressed, cannot be executed effectively for reasons that are not clear or cannot be removed. For the purpose of this book, we have identified five distinct (but related) areas in which stumbling blocks particularly come to the fore. These will be discussed in more detail at the end of this chapter.

- <u>Finance</u>, illustrated by the seemingly impossibility to find adequate funding (even though public and institutional money seems to be sloshing against the thresholds)
- <u>Talent</u>, illustrated by the inability to find people adequately trained to do the job. The reason is the apparent impossibility or slowness of educational systems to anticipate the need for specific capabilities
- <u>Vested interests</u>, illustrated by the many, often oblique ways in which people, stakeholders or institutions protect their habits and positions
- <u>Bureaucracy</u>, illustrated by organizational rigidity and institutional resistance (in collectives such as governments, institutions, industry) to change
- <u>Political will and support</u>, illustrated by a lack of political potency.

Whether these are the technically correct terms to identify the stumbling blocks is less relevant than the urgency to find effective ways to deal with them.

**Existential urgency**

*The urgency to aggressively address humanities' problems in changing its social and technological infrastructure (in a constructive way) is such that we must mobilize all the knowledge and insights we have. Otherwise there is a distinct possibility that humanity as a civilized society will perish in the disruptions caused by climate change.*

To succeed we need insights into the nature of the stumbling blocks and we must incorporate such insights into our action plans. In the next paragraphs we make an effort to provide a context within which the nature of stumbling blocks may be understood from the point of view of the creation of a reduced reality, the development of science and the technological progress that followed in its wake and insights from complexity.

## Reduced reality and science

A reality that is reduced to basic conditions that satisfy the existential needs for food, water, safety and procreation is a *sine qua non* for any natural system that humans consider to be *living*.[84] We will assume[85] that through the process of evolution all living systems[86] in nature are "equipped" with their own specific reduced reality[87] and with a unique capability to adapt that reality to changes in their environment.[88] Compared to the time it took evolution to equip[89] living systems that way (millions to hundreds of millions of years), the time span in which humans and their societies evolved is extremely short.[90]

> **Reducing reality: different approach in the east and the west**
> It is impossible for us humans to grasp all causal relations in the real world. To manage this impossibility, humans reduce reality to a level that they can handle. The West and the East have approached such reduction in fundamentally different ways. The West settled on boundary conditions and restricted a refined analysis to within such boundaries, to then secure precision and reproducibility (of the analyses) within those boundaries. This approach attempts to increase the "knowns" while banishing the "unknowns" to caveats and background assumptions. Conversely, the Eastern approach reduced complexity to a limited number of patterns for transitions of states in general forms based on extensive accumulated experience. This sacrifices precision and reproducibility to prioritize the inclusion of "unknowns" by allowing ambiguity to remain.
>
> Adapted from the abstract of Atsushi Iriki's talk at the Para Limes conference "East of West, West of East", October 2016 in Singapore, https://www.paralimes.org/wp-content/uploads/2018/03/Atsushi-Iriki_bio.pdf.

---

84 For a more fundamental discussion of living systems, see https://doi.org/10.1098/rsfs.2018.0023.

85 Making such an assumption is, in itself, an example of creating a reduced reality.

86 The term "living systems" in this text refer both to the individual "living creatures", the population of such creatures and the systems that embody the reduced reality for such populations.

87 Like beavers who create a reduced reality by building dams that form lakes that provide them with food, water, protection against predators and a place to build a lodge to bring up their young.

88 All living systems that over time did not manage to adapt to such changes went extinct.

89 A "tooling" force in the evolution of living systems is the competition for some kind of limited resource, such as food, space, energy, power or wealth.

90 For an example, see *Sydney Brenner's 10-on-10: The Chronicles of Evolution.* Eds Shuzhen Sim and Benjamin Seet. New Jersey: World Scientific, 2018. https://www.worldscientific.com/worldscibooks/10.1142/as001.

In that relatively fast evolution, reality is not reduced to fit the slow changes in the environments. Instead, through their adaptability and inventivity humans play a progressively important role in adapting their environment or physical world to fit their needs, thus creating a new reality within the physical boundaries of our planet. Science, which is a methodological activity by which we try to understand our world, was (and is) also an important factor in triggering changes and adaptations.

## Science and complexity

Science has, in the last 400+ years, essentially been reductionist. It assumes that we *can* learn the workings of a complex system by taking it apart and examine its components in isolation. We *cannot*, because, if taken apart, complex systems lose precisely the character that makes them complex, namely the features and properties that emerge when components come together and interact.

**Limitation of disciplinary science**

*As disciplinary science developed and led to its great inventions, it also changed the systems we live in, in ways that were not intended, predicted or understood. So, while we were able to apply technology to different aspects of our life, be it our health, our food, our mobility, our use of energy, our entertainment, our safety, our communication, our trading arrangements or our wars, we do not understand its effects on the evolution, growth, degeneration and death of the systems we depend on. The focus on single components of a system did not lead to theories, concepts and insights necessary for understanding the system as a whole.*

In other words, we can **de**construct the physical world to basic laws and elementary particles, but we cannot (**re**)construct the world we live in from those laws and particles. In his iconic 1972 article "More is Different", Philip Anderson explained why: "As we put particles and laws together, new properties and features emerge that cannot be predicted from the laws and particles itself".[91]

---

[91] Anderson, P.W. *Science*, New Series, Vol. 177, No. 4047 (Aug. 4, 1972), pp. 393–396.

Most of the "progress"[92] of the last 400+ years is directly related to the capabilities of humans to study individual components in complex systems, to find the physical laws that describe their behavior and to establish linear relationships between cause and effect.[93] This "disciplinary" science led to technologies that (now) support or dominate our lives and societies but did not lead to original insights in the complex nature of our world. In fact, it is the other way around: to find physical laws, science must create a "reduced" reality to keep the complexity,[94] that obfuscates those laws, outside.

## Complexity and stumbling blocks

Many of the major problems that we face in the real world[95] stem directly from ignoring complexity from the start and creating a reduced reality (instead). Such "simplification" is largely without consequences as long as we just

> **Complex adaptive systems**
> One might say that complex problems are manifestations of the behavior of complex adaptive systems. Such systems have a large number of components, often called agents, that interact and adapt. Because of that, the behaviour of the system cannot be predicted from or "reduced to" simple cause and effect relationships.
>
> Of complex adaptive systems, it is known that any outside intervention leads to unexpected consequences. Thus, by their very nature, problems arising from such systems do not have solutions. But there may be ways to influence their behavior.

---

[92] Progress as the result of a social contract (originated in the Enlightenment) in which scientists were supposed to understand nature and its laws in order to control them for the betterment of humanity.

[93] Many of the linear relationships between cause and effect that we work with in our daily lives, are in reality first order approximations (simplifications) of nonlinear relationships.

[94] The 1948 article in the American Scientist "Science and Complexity" by Warren Weaver describes the progress of science by the nature of the problems it has been able to address. Weaver distinguishes three types of problems: Simplistic problems that are the subject of Newtonian science, problems of disorganized complexity that are the subject of statistical physics, and problems of organized complexity that are essentially the subject of complexity science.

[95] The *real* world refers to the world that we see as humans, the humanized world.

analyze a problem and treat it in isolation. However, it is problematic if we want to deal with it in a broader context. Then understanding of the dynamics and nonlinearities of the *complex adaptive systems* that lie at its root is needed.

*To proactively address big challenges in the real world we need to take their nonlinearity into account at start.* But the reality is (at least in our western societies) that all our analytic, educational, and intellectual problem-solving power is based on and structured along linear cause and effect relationships. In this way stumbling blocks emerge. We run into them while we try to address and meet the great challenges that humanity faces.

Complexity, with its embedded nonlinearity, needs to be dealt with at the analyzing stage of problems. Moving towards solutions has always failed when introducing complexity at the solution stage only.

**Evolution and complexity**
*From the point in time that the big bang occurred, the laws of physics and complexity have been features of the evolving universe. The laws of physics **allow** us to make visible the history of the universe all the way to the big bang (reductionism allows us to look beyond the innumerable properties and features that emerged in the ever more diverse and complex universe). The characteristics of complexity such as emergence, adaptation and self-organization **prevent** us from knowing the future (we cannot know what new properties and features will emerge as the universe and all that is part of it continues to evolve).*

A key to understanding the origin of the stumbling blocks that we encounter when we try to deal with complex problems proactively, is that in the western world we tend to look at a challenge in a way that presupposes that we can deal with it in a linear way.[96] Amongst others that imples that we can make predictions by thoroughy analysing the past and extrapolating the findings into the future.[97] Our system of organizing

---

[96]   As a result of the reductionism that shaped it in the last 400 years.

[97]   Another way to put this is that Western philosophy of science has been based entirely on the belief that nature is fundamentally simple, and any appearance of complexity is our fault — the result of incomplete information. Daniel R. Brooks (2020), *The Major Metaphors of Evolution*, page 9, Springer.

governance and governing, of finding information, of using knowledge (and passing it on) or of engineering a solution, is based on that premise.

Instead of looking into the true nature of a problem or challenge, governments, bureaucracies and big organizations tend to look at a simplification that allows them to create an idea of the size or the problem[98] and then solve it by balancing that perception with a linear intervention. Often that intervention takes the shape of pushing money in the system, money that we suppose can solve the issue.[99] But rather than solve it, just throwing money at it aggravates the problem. Money *interferes* with the complex system and causes the emergence of unexpected properties (or unanticipated consequences). With complexity as the core characteristic of the systems that we are dealing with, stumbling blocks may be identified as such emergent properties.

> **Interference**
> *"The world is governed by letting things take their course. It cannot be governed by interfering"*
>
> From Chapter 48 of the Tao Te Chin (ca. 2300 BCE) (translation by Gia Fu Feng and Jane English (1972).

## Five stumbling blocks

Being emergent properties of the same complex system, it is clear that stumbling blocks are connected. Yet the different ways in which they manifest themselves can be matched to factors that play a key role in the ways we manage our societies (also huge complex adaptive systems): finances (money), talent (education), bureaucracies (organizations), vested interests (stakeholders) and political will and support (politics).

The following paragraphs illustrate how these factors translate in stumbling blocks once we have to deal with proactive action plans.

---

[98] Understanding the "size" of a problem is part of the enigma of "predicting" the emerging properties of a complex adaptive system.

[99] Often, if that does not solve the problem (because of the discrepancy between the nonlinearity of the systems and the linear ways we deal with them), the amount of money is simply raised.

*Finances*: Governments tend to think in terms of money when it comes to dealing with problems. The bigger the challenge, the more money is spent on it. But dealing with the consequences of a disaster that **has** happened or is happening by throwing money at it, fails. This is especially so when such a disaster followed a long period in which the complex nature of the processes that led up to it was ignored (as is the case with climate change). When we want to proactively mitigate the consequences of disasters that **have not** happened yet, such processes cannot be ignored. But, even though far less costly than dealing with the consequences of the disaster, those processes do not qualify for government money because they do not fit the linear cause and effect regime that characterizes the thinking of governing bodies.[100] A perfect example are floodings. Generally, after the water is gone, the insight for the need of preventions has also evaporated.

*Talents*: To proactively mitigate the consequences of disasters requires understanding of the nature of the complex processes that (may) lead to a disaster as well as an intimate knowledge of the (local or regional) setting in which the consequences of the disaster will be felt. Few professionals that can be called upon, understand the nonlinear behavior of complex systems. And there is an enormous shortage of educators who can educate a next generation about complex systems. So, education and the availability of talent becomes a stumbling block once we develop proactive action plans to deal with potential disasters.

> **People and complexity**
>
> *Although they do not realize it or call it complexity, most people at a local level deal with it (manage it) all the time. E.g. Mothers who bring up their children in a setting that continuously changes as the children grow up, develop, and have accidents, while the father moves in and out of the family as his work demands, are great complexity managers. Mothers do not call it complexity, for them it is a natural way of life. Raising a next generation of professionals (starting at kindergartens), mobilizing local people, and enriching their world with the knowledge of professionals may be a way to go.*

---

[100] Like governments (national and international) and national and international financial organizations.

*Bureaucracy*: Bureaucracy is directly related to the idea that organizations (societies) can be governed by splitting them up in "components" and dealing with those in separate units (silos). As problems grow more complex, units split up in sub-units, increasing organizational inflexibility. As disasters tend not to unfold along the lines of such organizational rigidity, the lack of effective coordination and communication under clear leadership poses a phenomenal stumbling block for proactively dealing with their consequences.

> **People and disasters**
> *Humans have time and again shown to possess tremendous creativity in dealing with the consequences of disasters and often have embedded lessons learned from those in ways to be proactive with regard to dealing with disasters to come. For example, the lessons Singapore learned in dealing with the SARS epidemic translated in an effective way to deal with the Covid-19 crisis. Similarly, the inhabitants of the North Sentinel Island (part of the Andaman Islands) survived the tsunami of 2004, because their collective memory recognized the signs of the coming flood.*

*Vested interests*: Four hundred plus years of organizing society and its institutions in silos has resulted in historically embedded interests to keep such organizations as they are, far away from the complexity that those silos are designed to ignore. Thus, changing that structure of silos and the policies that protect and maintain them in the face of possible disasters will run into opposition that goes far beyond the individual who sees his position threatened. In nature evolution tends to bypass such stumbling blocks. However, that takes much more time than we have available to take away the existential threats.

*Political will and support (political potency)*: Politicians and the systems of government in which they operate are typically not able[101] to think and act proactively in the face of potential man-induced disruptions and their consequences for society. This inability is exacerbated by a lack of understanding of the complex causes for these disruptions, making it impossible to translate *the fact that the consequences of disasters can be mitigated with far lower costs to society if proactive planning is applied,*

---

[101] E.g. Due to election cycles that punish long term vision over short term visibility.

in *political agendas*. This stumbling block is sometimes referred to as the collective action trap.[102]

Too often when comparing longer term future scenarios with and without effective action to address a disaster or adjustment of policy, the short-term financial and political costs seem more imposing than the long-term societal benefits.

### Stumbling blocks and collectives

One aspect of stumbling blocks must be discussed separately. That is their close connection to collectives, groups of entities that share at least one common issue or interest, or work together to achieve a common objective.[103] Looking at the list of organizations[104] that were mentioned during the discussions at the webinar (within the context of collectives), one gets an idea about the specter of objectives for which collectives are formed.

### Context and perspective

It is true for all collectives that, when they were set up or appeared, they were seen as (the way to) the solution of some problem larger than can be solved at a kitchen table. Collectives are necessary to get things done. But when the problem is solved, or when the environment changes, and the collective loses (even part of) its purpose, what was seen as a solution becomes a problem. Such problems then

---

[102] Collective action traps refer to situations in which the members of a collective pursue individual profit or satisfaction, rather than behave in the group's best long-term interest.

[103] For more information about collective action traps, see https://en.wikipedia.org/wiki/Collective_action_problem.

[104] In random order: Shell, British Petroleum, HSBC, World Bank, Central Banks, International Finance Corporation, Danone, Unilever, Walmart, Amazon, the Military-Industrial complex, Coca-Cola, religious organizations, multinational corporations or power groups that own, control or manage resources like fossil fuels, technology, finance or media, each with their own, secondary, objectives.

transform into stumbling blocks. So, directly or indirectly, a major source of stumbling blocks turns out to be collectives.

If we want such collectives to change themselves towards policies, decisions or a modus operandi that favor a transition to a world that can cope with the existential threats of climate-change and other human-induced burdens on the environment, we must find a way to break into their reality and create a movement of change from the inside. But, as indicated in the next paragraph, it is hard work to get into these collectives. To go around them instead looks logical and attractive, but the power of big collectives that have outlived their usefulness cannot be ignored. They will fight to survive with all the means at their disposal, which includes anticipating and blocking ways around them. Attacking such collectives directly is a waste of time and energy. Yet, if we wait for them to fail by themselves, they will take everything down, even those things we wished to remain.

This sharply positions one of the problems we face in dealing with collectives. We cannot ignore them, yet we cannot fight them directly. We need to find ways around them.

Finally, to deal with collectives, new ones need to be set up. By their very nature these too will outlive their usefulness and then become a problem. We need to recognize why and when they can become a problem and how we can avoid or minimize their negative impacts.

## Characteristics of collectives

Collectives are resistant to change. They are often old because they have incredible survival skills. They can be large or small organisations knitted together very tightly in *the fabric of society*. They have a hive mind[105] and they have complicated ways to behave. Increasingly, they do not behave like people want or expect. Though the members

---

[105] From the Oxford English Dictionary (https://www.lexico.com/definition/ hive_mind): "*A notional entity consisting of a large number of people who share their knowledge or opinions with one another, regarded as producing either uncritical conformity or collective intelligence*". Alternative definition: "*Collective brain*".

of human collectives are people, the collectives are not human. And the bigger and more global collectives become, the less human their behavior. Nonetheless, they do have a common feature with other living organisms. *They want to live. They will not commit suicide voluntarily.*

One of the largest, oldest collectives on the planet is the Roman Catholic Church. Some multinational companies are examples of enormous contemporaneous collectives with great survival skills. Many collectives create a feeling of belonging and an environment of like-mindedness for those belonging to them. But by their sheer size, these collectives have considerable investment and intangible infra-structure, such as personal networks, relations that are conservative and highly resistant to transitions such as sharing property rights.

> **Using collectives**
>
> *As indicated already, we cannot ignore collectives and we cannot fight them expecting to win. But collectives can also play an important role in the development of ways to address threats to our existence. Multinationals for instance, have well developed distribution and collection networks, whether in data, resources, products, or money. Not using these may be double edged. We should therefore look for ways to align all potential players in finding ways to cope with the big threats while going around the stumbling blocks imposed by collectives.*

All the big structural decisions of the world are made by collectives, not by people. People make decisions within collective space but those decisions are constrained by the conservative properties of the collective.

Decisions at the global level are increasingly about the biggest collectives, that often behave in very different ways from what happens at local levels. That points to what Daniel C. Dennett coined as *competence without comprehension,*[106] an evolutionary principle that roughly says: *more intelligence arises as an effect of less intelligence,* not the other way around.[107] Thus, handling complex situations may be better served

---

[106] For more information on the issue, see https://www.youtube.com/watch?v=m6BvEtWBErQ and https://adamsmithesq.com/2012/07/competence-without-comprehension/.

[107] Translated to the discussion about collectives: *We have created something that is so complicated that very few people, if any, actually comprehend what it is.*

by bottom-up processes at local levels than by top-down decisions through collectives.

By their very nature, survival skills and size, big collectives present enormous barriers to change. They will not change themselves unless their survival is at stake. To survive they generate stumbling blocks of which the protection of vested interests may be the most dominant. But finance, bureaucracy, political potency, enabling policies, keeping a tight hold on their talent as well as using the media to their advantage, are very visible too. And, as we mentioned in Chapter 1 (textblock *Merchants of Doubt*) sowing doubts is a powerful tool to prevent decent policy-making, despite clear evidence of facts.

The different projects, explorations and approaches that are described in Section 2 of this book will show that it is not always clear how to go around collectives or avoid stumbling blocks or how to include collectives in projects or efforts aimed at thwarting existential risks to humanity. At the same time, the different approaches that are discussed and described implicitly present a broad spectrum of experiences by leaders in the world to do just that.

# 12   Selected discussion highlights

The **purpose** of the webinar was to:

- identify major stumbling blocks that emerge on the way to buy time for developing action plans to mitigate consequences of global climate change
- explore ways around those stumbling blocks using grassroots initiatives

### How does the current situation differ from other instances in which we have had to problem solve and how have we solved similar problems before?

Human history tells us that our biological response to disasters is to run away. But we have now become a majority (highly sedentary) urbanized species and we have taken away that option (or at least it is not an option that is seen as easy). This means that we must find a different way to follow our biological impulse to flee from disasters.

> In framing our assertions that we are dealing with an unusually urgent problem, we need to include more information from biological history, anthropology and human history.

### How can we convince those who have the power to say yes that proactive measures are cost-effective?

Three key notions are:

o Business as usual is not working
o The cost of inaction is unacceptable
o Resisting change is not an option, or not going to stop climate change.

*In our action plans we need to make plausible arguments for cost-effectiveness. We need to use the language of prevention rather than crisis response.*

**What is cheaper**, coronary bypass surgery or diet and exercise? Which is more effective? This is well understood for non-infectious disease but not for infectious disease for some reason.

This is not taken seriously in discussions about action plans (such as managed retreat) associated with global climate change. How can we create a sense of urgency before crisis response is necessary? We need to have a sense of urgency that is sufficient to catalyze voters, and to catalyze cooperation.

## Getting the words right

- Think the **unpalatable**, not the unthinkable.
- Replace the (false) dichotomies of optimism/pessimism and positive/negative with **realism/being realistic**.
- It is not climate *change* we should talk about but **climate *emergency***
- Talk about **political *nerve***, not political *will*. The **nerve** to tell people what they need to hear rather than what they want to hear, to tell people the bad news, and to assert a bold vision for effective action, which means being proactive.
- People of heterodox ideas are often called "rebels, or mavericks", they should be called "**visionaries**".
- **The New Normal**: actions to begin to cope must begin now and must continue into the foreseeable future.

## Getting the reference frames right

- Funds and financing are available, just as there is enough water, but the distribution is screwed up. Funds and financing are there but they are directed from one crisis to the next — for example, funding for managed retreat comes prescribed for only certain kinds of measures and is not fungible to other areas (like hinterland regeneration). Funding and finance must be re-directed into "preparedness and prevention".
- We need to take a **complex-systems view** and especially one in which we eliminate the traditional top-down/bottom-up

dichotomy. Putting this into practice means that we need to identify capacity in local situations and nurture it upwards. This means creating minimalist bureaucratic scaffolding for those with unusual capacity at the grassroots to make themselves and their ideas known higher up. The role of top-down infrastructure is to be sensitive enough to unanticipated innovations that come from the grassroots and provide the tools to capitalize on them. Since global government is impossible, we have to work at efforts at the local level to try different methods of adaptation. Trial and error will help different communities to find the best path to fit their need to adapt and to mitigate the consequences of climate change.

> *Implementing* this will require particular kinds of personal commitment — a mixture of a desire for public service and high-quality research. Examples of this are the work with parataxonomists in Costa Rica, and VectorAnalytica in Latin America.

o One needs to recognise that **one size does not fit all**, that context is critical and that best practices are not always transferrable. Do not *impose* capacity-building, if "grassroots" is simply seen as the local extension of a top-down agency. This is the problem with organisations like *One Health* and *EcoHealth*.[108]

o Corporate entities need to join donors and NGOs and shift emphasis from shareholders to stakeholders. We need to emphasize that change is inevitable, but change does not have to be bad and resisting change may make things worse.

> *Change*, in the form of migration in some form or another, has often been driven by simply fleeing a bad situation, but is more productive if stimulated by the suggestion of better opportunities elsewhere.

o We need to avoid notions of "we are helping [the other]", "we need to take care of them" perspective. This creates a context of making people dependent; human rights have to be comprehensive. (This is about all of us, not us versus them.)

---

[108] For more information, see https://www.cdc.gov/onehealth/basics/index.html and https://www.ecohealthalliance.org.

o Emphasis for managed retreat: We should not confuse equity with equality.

## *Water and money*

Water and finance are vital elements to human survival and civilization, with stock and flow implications. Water is physical, money is mental, but its derivative is changing all the time. Water is available and, like money, it is available at global levels, but it is not distributed evenly. We do have the technology to handle the scarcity. Physical money like gold is limited, but fiat and cyber-money[109] like bitcoin[110] can be created at will.

Water, finance and climate change are interrelated and entangled issues, but each is managed in a silo and those silos do not talk to each other or have difficulties communicating with each other. Trust has been lost because each (specialist) silo — call them "elites" or "power clusters" or "power networks" — overpromises and underdelivers. They present a solution in an ideal situation, but when it comes to policymaking, they always underdeliver, and trust has been lost because of that. (Technological or ideological optimism without accounting for the dark side.) We need to realize that all change will require very hard trade-offs and there is a danger of trying to promise too much and not talk about the costs that are involved.

---

[109] **Fiat money** is legal tender whose value is tied to a government-issued currency, like the US dollar. **Cryptocurrency** is a digital asset that derives its value from its native blockchain. The issuance and governance of fiat currency are dictated by central banks, while blockchain protocols, code, and communities govern cryptocurrency, https://www.gemini.com/cryptopedia/fiat-vs-crypto-digital-currencies.

[110] **The mining of bitcoin** consumes more energy than a small country (like Sweden or the Netherlands), https://digiconomist.net/bitcoin-energy-consumption/.

The reason why money is always pumped out into a crisis is because it is the illiquidity or excess liquidity that kills all transactions and causes a complete break-down, which is an important issue that mainstream economists have always ignored. Like water, scarcity of money creates drought, while excess money creates flooding. They have systemic consequences, which is why they are public goods and usually con-trolled by law. Time and time frames affect values and decision systems, and time scale issues are increasingly critical in our understanding of the change element of all this, because change occurs over time.

Both water and finance carry emotional, religious, and psycholog-ical characteristics, so rational or market solutions may not work at mass or sectoral levels. Trust, transparency and values are deep issues that must be addressed, among others through the "happiness" approach, as one possible way to deal with this high complexity.

> **Insight contributed by Martin Rees**
> *"Water and money flow, liquidity, is critical to human activities, but these operate in different time cycles and scales. If they do not synchronize, it leads to disruption, breakdown, or collapse. So we can really see the whole ecosystem as different feedback cycles, and these cycles create change".*
>
> This insight is increasingly unclear in academia. Money has simultaneous asset/ liability/ trust characteristic(s), that facilitates trade and transactions.

## *Barriers, property rights and pipelines*

The **barriers** to change are mind-sets — which have been influ-enced by our evolutionary past — and big collectives like reli-gious organizations, multina-tionals, power groups, that own, control or manage resources like fossil fuels, technology, finance or media, each with their secondary objectives. Sometimes there is the primary objective of survival,

> **Lessons**
> *There are important lessons to be learned from the success and failure stories in water and financial management at local, national, regional and global level. But lessons are not widely understood or shared. For example:*
> - *The key institutions in finance are central banks, who are at difficult cross-roads on supporting elites or masses.*
> - *Because of corruption, knowledge and funds are not allocated where they are really needed.*
> - *The synchronization of information is absolutely critical.*

which may hold out, prevent, corrupt or capture policies that attempt to deal with climate change or progress, because they feel they are threatened. Many of these collectives create a feeling of belonging and create an environment of like-mindedness for those belonging to them. But by their sheer size, these collectives have considerable investment and intangible infrastructure, such as personal networks, relations that are extremely conservative and are highly resistant to transitions and sharing **property rights**. You cannot move into climate change without reallocating property rights, so there are losses, and the question is: **how do we allocate these losses**?

Inadequate or non-existent property rights (this includes access to resources) prevent investment in sustainable solutions for the future. Property rights are complicated because everyone thinks you can trade these rights. However, common rights are difficult to trade, yet very easy to expropriate. Recent experience reveals that technology, money and local knowledge is available, but local solutions may interfere with national strategies and interests, or there is no expertise available to design or manage complex projects, leading to a lack of concrete projects. This is a **pipeline** issue. Even when there is money to be given for all these projects, there are no projects in the pipeline to use this money. These pipeline issues can be solved quickly to foster trust at the local level but delays in getting projects in the pipeline and the wastage associated with these hurried projects create distrust.

> **Resource sharing**
> *Responsible resource-sharing as we know from local fisheries can be extended to terrestrial wildlife for meat, preventing negative impacts on the environment by huge cattle farming. If done right, this is a positive way of managing biodiversity, protecting the environment, limiting the negative aspects of the race to the bottom and thus preventing the tragedy of the commons.*

## Building trust

Trust must come from the bottom-up. For that to happen, leadership must be transparently accountable and deliver on their promises. Building trust can be facilitated by ethical leaders and academic and other research specialists who are willing to combine social service with love of research (such people must be rewarded for those efforts); we need competent compassionate leaders to respond to what comes up from the grassroots. This is not something new. We have known about this already since Shakespeare.

> *We live in a time of profound fear*
> – *when no one trusts leaders, that no one trusts data, or the people (scientists) that analyze them.*
> – *when we throw money into activities that are not working rather than take a chance on innovation.*
> – *of paradoxes that politicians and scientists are unable to explain*

> **Fear** *of death* **makes us all cowards,** *and our natural boldness becomes weak with too much thinking. Actions that should be carried out at once get misdirected, and stop being actions at all.*
>
> — Hamlet, Act 3, Scene 1 (1599–1601)

There must be real respect for the people involved, otherwise they will not trust and the attempt to build trust will fail. We must build cooperation, and we need to know that we will (and must) cooperate with people that we do not like. How to do so? We will have to deal with issues that are border busters — e.g., pathogens do not recognize national boundaries.

> *Many solutions will ultimately cross political boundaries (and in some cases, these are easier for officials to invest in than purely internal conflicts).*

## Food and the food system

o People that deal with food, fish, basically have their feet in the dirt.

o The current system of global food supply is based on production/consumption growth, not on consumer and farmer/ecosystems well-being and health.

o There are major differences between the developed world and its problems and the developing world. We need different stories for both areas. Action is needed. Local actions can help on a limited scale but will not be sufficient. To make changes possible, you also need global or national action. Combining local and systemic change is critical and requires agents of change. How to find and support them?

o We cannot change the system in the short term, so whatever is going to happen will be step-by-step, (sometimes baby steps) in many areas rather than broad developments that change the system.

o Action could be more effective if we look at adjustment of the funding of the banks and trade regulations, and internalizing externalities. That is, how do we measure the value and the real costs of what we are doing.

o In the developing world, local food production faces extreme risk factors. If we want to make changes in the developing world, we must look at reducing those risk factors. From a finance and systems point of view, that is something that we should focus on much more than what we have been doing in the past.

o Certain collectives have not been well-defined, but it is important to get certain people, who play an important role in the food supply, in one room and say: "Okay what are the things at this point and in the longer future, that we can deal with?" In addition to that, there is advice necessary at various different levels, where we can push the food system to adjust. Worldwide is as important at the local level or even at the local fishing village.

## Mental maps that lead to collective action traps and key challenges

There are at least six mental maps that are not in sync with each other, which is leading to collective action traps.[111]

1. *Humans still behave in a tribal manner and are very tribal centric.* While that has made our species evolutionarily successful, we are now faced with global issues that we never faced before, and which cannot be solved with the tribal approach.

---

[111] A collective action trap, or collective action problem or social dilemma is a situation in which all individuals would be better off cooperating but fail to do so because of conflicting interests between individuals that discourage joint action. For more information, see https://en.wikipedia.org/wiki/Collective_action_problem.

2. *A mainstream economic paradigm that dominates policy formulation.* In that paradigm everything must be reduced quantitatively to money so that we can do trade-offs. But this is partial, equilibrium-centered, zero-sum, market-oriented, and it ignores ethics and morality and does not properly account for ecological costs. There is no accounting for externalities, biodiversity, resource depletion, pollution, destruction, etc.

3. *Big collectives are very conservative and want to maintain the status quo.* These power elites deny the need for change or anything that is against their interests, because they see this as a loss of power. This is often coupled with corruption etc.

4. *A science community that is very open-minded and produces new technical ideas yet remains locked in specialized silos that are not connected to the center of climate change policymaking.*

5. *The young generation (like Greta Thunberg) cares about climate change but does not know how to make the necessary changes.* They need to make their walk through the institutions, which is where grey-haired people like us may play a role.

6. *System view of life,* where a change happens organically and holistically, with two possibilities for a human role — no action (go with the flow or market self-order) or action-oriented (state, private or civil society led). Not either-or, not zero-sum, but all entangled with system emergence. The "happiness" approach may be a way out.

## What are the key challenges?

A. How do we switch from limited scale tribal behaviors to one global tribalism? The challenge is to find mechanisms to steer humans towards more sustainable, inclusive and global solutions. We are all in it together. **Man, and nature are one**.

B. If water, money, and technology are not constraints to buying time for climate change action, what specifically can this group[112] do to push for change?

---

[112] I.e. the participants of this webinar. For the list of participants, please refer to Annex 1.

C. What is the imaginative idea that would make a difference on global change?

## *Additional thoughts*

o The key issues are systemic, complex, and not amenable to simple solutions. No one-size-fits-all solution is possible. But identification of mental barriers enables us to seek a common framing of issues and challenges, identifying tools, concepts, funds, technologies, and values that may lead to alignment of incentives and goals to "nudge" change. Acceptance that we are all in one planet and must cooperate to survive or thrive may be an important direction.

o No individual or polymath can now grasp the massive growth in knowledge and technical expertise, or the emergence or evolution of giant open complex systems. Things that are just changing too fast or complicated for any individual to solve.

o Different silos are not cooperating with each other, but it is better at scientific level.

o If we can get an AI analysis of clusters of solutions that may be cheap and (relatively) easy to implement at local levels, then we avoid being caught at Big Power politics.

o We really need to get out of the mindset that "only the market can solve these problems", this zero-sum, either-or. We have been very much in a monoculture, and that has created a fragility whereas history has told us that system diversity, greater biodiversity, allowing a thousand flowers to bloom may be the solution to go.

o Coleen Vogel has drawn attention to the platform "Humanity Rising[113]", formed by a group of people who are passionate about bringing communities together to empower young people to change the world.

---

[113] "Humanity Rising" will open to the public in the future. For more information, see https://humanityrising.org/about-us/.

### Will inaction lead us to begin talking about "triage"?

- When do we have to make decisions about which ongoing activity is worth funding?
- How much time do we have before we replace incentives with coercion or simply abandon those who will not cooperate?
- How much effort do we put into people who do not want to change even at the potential risk to their lives? (For example, when do we abandon climate change deniers to their fate?)
- We may have moved to the stage when we must concentrate on mitigation and adaptation, where we accept triage and try to get as many people to focus attention on cooperation and shared action, rather than fighting like alphas in a burning forest.

## Priorities of change to buy time?

1. Let multi-background teams — largely independent of existing bureaucracies — supported by scientists, translate existing global climate change projections into transparent (multiple) national and sector long-term scenarios to guide future activity programs to address climate change and enable identification and ways around stumbling blocks

2. Identify and support people with proven high-level leadership qualities and politicians with nerve to direct adjustment of existing macro-policies and regulations and enable regular, transparent monitoring of their implementation and effects to instill integrity, trust and respect in this complex political process.

3. Expand education and research capacity to develop human capital in private and public sectors that are able to function in complex environments and to redefine the prevailing economic narrative to implement climate related civic, public and private/corporate policies, regulations and actions.

4. Encourage private and public exposure of vested interests opposing change, and actively support the definition of policy and particularly private action to circumvent their opposition effectiveness.

5. Support global efforts and create national macro-policies, consumer and political support to better facilitate fast and fundamental adjustment of critical industries and sectors, such as agriculture, trade and urban development.

6. Define and create flexible systems to expand financing climate related adjustment at global, national and particularly local levels, better reflecting the transparent and fair distribution of risks over time and reducing the window of the collective action trap.

7. Incorporate major externalities and costs of no/insufficient action in analyzing and recommending climate related policies, market regulations, costs and prices, and encourage a pro-active approach whenever possible.

8. Focus climate related strategies on the critical duality of the role of high level (national and global) policies, while providing maximum flexibility for the sharing of power and implementation of multiple, parallel actions at lower levels aimed at converting unsustainable production, trade and human settlement practices.

9. Exploit unavoidable crises or wrong-headed policies as levers to circumvent implementation obstructions and welcome drivers to accelerate change.

10. Accept that fairness should not only be perceived but also proven for current and future generations in all policy and on the ground action as essential basis to build public trust and reach tipping points to modify public attitudes.

# Section 2: Exploring ways around stumbling blocks

*New beginnings are often disguised as painful endings* — Lao Tzu

The explorations in the following chapters represent the fruits of the discussions during the webinar: *Removing barriers to buy time*, that took place in February 2021.

These discussions were carried out in three small groups. They were triggered by the presentations that are reported in Chapters 3–9. They then moved from state-of-the-art insights in the development of climate change and its consequences, to (potential) problems related to it, to ways to address those problems and to philosophy, all in rather random order. The intent was to identify and discuss ways to deal with stumbling blocks that hamper the execution of action plans to buy time to deal with the consequences of climate change. **One common thread that emerges from the narratives is that it is naïve to believe that we can eliminate the disruptive power of stumbling blocks by attacking them directly.**

After analyzing the separate discussions, it became clear that **collectives** played a central role in all discussions, and that one subtext of the discussions was that **the way to deal with collectives is to go around them.**

# 13  Introduction to section 2

Society evolves in unpredictable ways. If living systems (like organizations) are to sustain themselves in that society, they must evolve as well. This means that they must be able to **exploit** local conditions sufficiently to survive **and** be able to **explore** the boundaries of their existence to cope with changing conditions.

Living systems are exploitation-biased during times of relative stability, and exploration-biased during times of change. We now live in a time of accelerated change. To cope with that change, exploration must prevail over exploitation. There is no way to know to what new conditions changes will lead, so the greater the potential for exploration the better.

Organizations will have to shift their focus from maintaining the status quo (like protecting vested interests) to exploring ways to cope with the changes in society.

Looking for ways **around** the stumbling blocks that emerge from the system must prevail over efforts to change the system that generates those stumbling blocks. That is what Section 2 of the book is about.

That is, in fact, what this whole book is about: **exploring ways around stumbling blocks**.

But we need to warn you, our reader, Section 2 is not a smooth narrative or a systematic treatment of ways to get around the stumbling blocks. Rather, it is a narrative of narratives, told from the perspectives of different experts (leaders in their fields) and their experiences in areas where time is in short supply when dealing with the consequences of climate change. One common thread that emerges from these narratives is this: it is naive to believe that we can eliminate the disruptive power of stumbling blocks by attacking them directly.

Section 2 is a collection of dialogues that together show an emerging state-of-the-art in dealing with stumbling blocks (without mentioning them explicitly). Some of the stories originate with a problem, like shutting down a coal fired power plant. Others start with a philosophy of life, like the Balinese Tri Hita Karana, that centers around (building and maintaining) harmonious relationships.

All the stories in this section are interconnected and many get their richness from details that link them up with other stories. That, in itself, is not surprising because all stories deal with complex systems one way or another and, as argued before, we cannot effectively deal with complex systems by reducing them to simple cause and effect relationships.

If you, the reader, feel passionate to help avoid an existential disaster for humanity, continue reading. For those restricted by time or interest, the last chapter of the Interlude (Chapter 12) contains some highlights of the rich discussions that are the substance of the following chapters.

# 14  Decarbonization

*Gary Dirks*

In the work that I do, we follow two broad directions: The **first** direction is to anticipate what kinds of tools, technologies, and insights will be needed as we move toward the middle of the century. The **second** direction is how do we make this go faster. And there, at least in the schools that I work in, it is important to recognize that a lot of the debate about the climate was substantially over the past five years. As one of my colleagues is fond of writing, the contest between solar and fossil fuels is over, and fossil fuels lost.

But that does not mean that we have a very good idea of exactly what to do next, and exactly how to go about doing these next couple of steps and doing them in ways that bring people with us. One of the criticisms I have had of the academy is we fought so hard to convince the world that climate change was a serious problem, and we fought so hard to win over some of the more difficult elements of society, but we did not spend enough time thinking about what we are going to do when we win.

And this is something I raise particularly with Non-Governmental Organizations (NGOs) that are active in the areas I am looking at. I think there is a very important role for groups that are well-positioned to facilitate genuine conversations about the next few steps, and how we are going to go about them.

## Shutting down the coal plant

We had the largest coal plant west of the Mississippi in northern Arizona. An enormous coal generation facility, on the order of a few gigawatts of coal power. There were huge campaigns to shut this down, particularly by some of the big interests. The effort to shut the plant down was resisted aggressively by the lead utility that owned the plant for a long time. There was a controversial study at Arizona

State University that said it would cost the economy billions to shut this plant down — in lost jobs, in lost economic opportunity.

Two years later, the company announced that it was shutting it down, in part because of pressure on emissions but also because solar was so cheap in the state — the plant was no longer economically viable. But embedded in that decision was also the reality that the Navajo nation would lose about 30% of their revenue, and the Hopi nation would lose about 80% of their revenue. Something on the order of a thousand jobs were going to be lost in a part of the state where there are no opportunities for new jobs. There was no plan.

> **Alexander Zehnder:**
> Indeed, it is the indigenous and local people suffering the most.

We held a conference after the events, and the city manager of this small city of about 5000 was there. Many of you know that Arizona has some very divided politics; we have some very conservative people in the rural areas, and he was one of those. In his closing remarks, he said, "I want you to know that I'm not looking for a handout. That's not who we are. But I also want you to know that we do not know what to do."

> **Peter Robertson:**
> Fossil fuels have lost, the big question is "what now?" And seeing that the sponsors of change are almost in a sort of shock — what now indeed.

Many NGOs keen to shut down the plant had no idea what to do either; their goal was to shut it down, full stop. I had a conversation with the head of the Sierra Club,[114] whom I talk to from time to time. He is a very good guy, very committed to decarbonization. And I was quizzing him a bit about what he planned on doing. But eventually he just said, "You know, Gary, I don't worry about what happens after we shut these things down, I just shut them down."

---

[114] A grassroots environmental organization. For more information, see https://www.sierraclub.org/about-sierra-club.

### Explore

I am raising all this is because I think (on decarbonization) we have the wind fully in our sails behind us. But people need a lot of help, including some of the main actors that we sometimes vilify. This company[115] that was fighting so hard to keep the plant running, we and they just announced a major strategic partnership about three weeks ago.[116] With a whole range of sustainability-related goals, they are now among the more aggressive in the southwest trying to decarbonise. But they need a lot of support. So, I think that there is a huge opportunity now, particular for the academy, and not just the academy, to be taking a second path — how do you actually make this stuff happen, and how do you make it at all scales, and in particular, how do you recognise that there are "collectives" — big and small organisations knitted together very tightly in the fabric of society. And they all need to figure out how to move from one convergence point to another, when they see it as risky (potentially an existential risk) and have no tools with which to do so.

> **Alexander Zehnder:**
> *If the locals are not taking up such fundamental discussions and building up pressure, little will happen.*

# 15   The financial conundrum

### Francis Vorhies: Pipelines, financing, shortage of small projects

In the early 2000s, I was involved in a project with the International Finance Corporation (IFC) to create a biodiversity business investment facility in Africa. We called it the *Kijani*, the Swahili word for "green." And it came out at one point, and this was about 2001, that the IFC decided they did not want to invest in any project in Africa that was below

> *There is a need for small-medium credits for projects on the ground.*

---

[115] The Salt River Project.
[116] January 2021.

30 million dollars. At the time, how many $30 million projects were there in Africa outside of mining and oil and gas? There was one giant aluminum smelter going up in Mozambique, and that was it. And what we were looking at was of course ecotourism, but also seaweed farming, and wildlife ranching, and we were looking at organic coffee, organic beef, all sorts of different landscape-friendly investments. The most they needed was maybe a million, a million-and-a-half. And there was absolutely no way that IFC could bring its focus down to landscape-level investments. Everything had to be much larger scale.

Maybe there is a real need to talk not about micro credit, but small-medium credit, and look at how we can move investments into landscape-level projects. They could be in timber, they could be in carbon, they could be in whatever, but they have to bring it down to a level that communities live at. We do not live at the level of BP and HSBC. Financing is all geared up there, and even well-intended financiers like the IFC are looking for projects that are too big to impact people's ability to use that money for livelihoods on the ground. So, a part of the solution, might be using modern technology to get financing down into smaller projects. And then people could take your water purification scheme if they want it, at the local level, because they could get the funding for it.

> **Define and create flexible systems to expand financing climate related adjustment at global, national and particularly local levels, better reflecting the transparent and fair distribution of risks over time and reducing the window of the collective action trap.**

### *Ambitions and reality*

One of the transits in the conservation world today is a campaign that says nature needs half. Or at least 30%. Boris Johnson in this country (the UK) the other day said he is going to set aside a third of this island for nature. There is a problem with that because it means zoning half of the planet as the other side and this side as the place where we hang out and get our water and our food and everything. "It looks to me that maybe we're bringing this down the wrong track." The track should rather be: how do we do sustainable land management?

Let me bring this back to finance — I was in a call last week with International Union for Conservation of Nature (IUCN). We were trying to put up a big nature-based solution investment fund. But the big problem in our sector is there is no pipeline, nothing to invest in, no investment opportunities.

Earlier I was on a call with United Nations Environment Program (UNEP) about un-locking the wildlife economy in Kenya.

> **Andrew Sheng: pipelines**
> *The issue about finance that is readily available, and lack of pipeline is an old one. When I went to the World Bank as a mid-term career rather than straight from university, the Young Turks — all perfect PhDs, no experience on the ground — took over. All macro. And the engineers, the project managers, they were all sidelined. The result: the developing countries had no projects. So you have a lot of money, and all the macro guys preached markets, markets, markets, and yet a ton of money cannot get any projects on the ground.*

We had a project with the ministry of wildlife and tourism there to open up beyond tourism to look at meat, at food security, carbon credits and so on. The ministry backed off and is nervous because there is so much pressure to set aside animals you do not touch and so on. So now what is happening in a place like Kenya is that livestock industrial agriculture has gone up, while wildlife use has gone down. You see cattle replacing kudu, replacing wildebeest.

So, if you want to connect landscapes with food and climate change, one big issue is the use and ownership rights to wild resources. Then wild resources

> **Use and property rights**
> *Filling the investment pipeline for sustainable land management (to connect landscapes with food and climate change) requires use and property rights.*

become part of the solution, not just something that is set aside, that you go look at on a holiday, but something that you actually utilize. So the discourse in Kenya is how can we develop the wild meat industry, to convert the landscapes into more sustainable climate-resilient landscapes; move out of cattle, move out of goats, move out of camels — which is a big part of it — and move into wildebeest, kudu, and the other biodiversity species that will be associated with a rewilding exercise. Hence, part of the issue is getting the use and property rights allocated to deliver sustainable outcomes and this relates to the issue of incentives. This does not mean a free for all, but it means identifying what package of rights will encourage landowners and communities to be able to use their resources sustainably, in a climate-friendly way and so on. And currently we do not have that package.

> **Comment by Andrew Sheng**
> *When you are talking commercializing property rights in a way that they are viable, you have to realize that the more you have human beings eating wild meat, the more you get pathogens — so there are issues there that are complex.*

I will give you another example and there is no one right answer here, only different answers. In the US, people are allowed to harvest wild meat but they cannot sell it, they can only bring it home and put it in the refrigerator. Which they do, and they take out tons of (non-endangered) animals. In North America, there are huge stocks of wildlife, the system works very well. In Africa as well, South Africa is the one that has huge stocks of wildlife, and in Namibia, because they do have a wildlife economy and there are use rights in communities, and private owners own them and so on. But in much of the world, we do not have the use rights well-allocated.

We know there is this issue in fisheries. It is the same thing with terrestrial too. So I think part of the solution is to look at the issues from a sustainable landscape, sustainable seascape perspective and say: how can the people that live there have rights to the resources, with those rights constrained that the uses are going to deliver sustainable outcomes that are climate-resilient, that are sustainably harvested, and so on.

In North America, one discussion is whether we can move to commercializing wild meat so it does not just go into the hunters' freezer, but actually goes into regular restaurants and grocery stores, creating an alternative to industrial livestock that we have in the country that has outcompeted wildlife because wildlife cannot get into the marketplace and cannot compete there.

So there is this thing of law, property rights, ownership rights, landscape-level, community engagement that would then create the potential for pipeline — an investable pipeline. Right now, we cannot invest because we do not have the right to use the resources in a way that would attract capital.

# 16 Wildlife management, fish and meat. What worked and what did not

*Francis Vorhies*

Wildlife is not just wildlife that walks around on four legs, it also swims, and so the first one that a lot of people are unfamiliar with is the management or mismanagement of wild fisheries. The major hunting operations on the planet are all offshore, and a lot of the industrialised hunting of fish is unsustainable. But there is a lot of discussion about getting artisanal fisheries to our local communities, locally-managed marine areas, and so on. And in some places these seem to be working better than others. Surprisingly, with my work in Yemen — and Yemen was very interesting in that it had an artisanal fishing industry — it kept out the big industrial guys and they managed the resources pretty well. The country has got other problems, but the fishery side sort-of worked. In a country like Namibia, over half the meat — now back to terrestrial meat — is wild. And it is wild-harvested in community-based conservation areas and private areas. And so people go into the stores and they buy wildebeest as opposed to chickens.

The alternative scenario you are getting is in places like Cameroon, where the US government is funding people to eat industrial chicken as opposed to game meat. And so what we are seeing is that in some places, the move away from use of native resources is causing ecological problems, causing diseases — here they have just put down a lot of industrialised chickens because of bird flu. Industrialised animals also get infected.

| *What works* | *What doesn't work* | *What is needed* |
| --- | --- | --- |
| *Artisanal* | *Industrial approach* | *Localized policies instead of industrial approach* |
| *Locally managed* | *Control away from local resources* | *People must have control over local resources and landscape* |
| *Keep out the big guys* | | *Get away from economic models and progress thinking* |

But I think what we can see is, in North America, where game meat is utilised, we do not have endangered species. Same for Southern Africa. It is in the parts of the world where they have said, "No, they have to be in that 30%," that the animals are disappearing. In Europe too. In England there is a very developed industry, but right now there is a bit too little consumption of venison. Most of the venison in that country goes into the restaurant industry, which has been closed by government policies because of Covid. So, the venison stock, which grows about 30% a year, is growing too fast this year — there are too many of them. So you could see policies saying, do we harvest here, do we stop here, and so on. And a lot of those policies really should be localised. It is important that people have control over their landscapes. It seems to be the way to work.

Some of it is national policy. Some of it in the European context might be at the EU level. But probably, there is more need for people to have control of the resources locally — fish, wildlife, timber, carbon credits, and what have you — and look at their landscapes as areas of responsibility, but also areas of

**Alexander Zehnder**

*It is not only wrong what multinationals do. Keeping multinationals out of the market severely hampered the introduction and distribution of **golden rice** a genetically modified rice with a high Vitamin A content that was developed 30 years ago. The researchers deliberately did not patent the rice because they wanted to make it available for everybody, particularly the poor who suffer from Vitamin A deficiency. About one third of the world population suffers from this deficiency which can lead to blindness. Now, after 30 years, the rice is still not on the market. And the reason is that nobody is pushing to bring this rice to the farmers and on the market.*

natural capital and wealth. And you can utilise those resources more effectively. That is the theme.

Given that we have these giant neo-colonial companies — the HSBCs, the BPs, Shells and the postcolonial World Bank that you cannot kill, that are part of the system and very powerful, the question is, can we decentralise control over our planet to a level of governance where people can live with it and benefit from it.

We have talked a lot about oil and gas, but you have the same problem with the food sector — the Danones, the Unilevers, and so on. They are giant beasts of companies (collectives) that are very divorced from local ownership and local control of people's backyards, people's landscapes. So it is governance decentralisation, I think, that is a piece of the solution, and using the resources that are there, instead of replacing them with industrialised resources that feed into what I guess you can call the Food-Industrial complex. So there is a similar problem.

# 17   Let a thousand flowers bloom

*Andrew Sheng*

Why are we in this bloody mess today? And how do we get out of it? What do we need to focus on? I think there is no one-size-fits-all solution. There is no magic wand to do this. We just experiment with many different types of approaches at the grassroots level, at the state level, at the corporate level, and the international level, and let a thousand flowers bloom, and who knows which one will take off.

I have worked with Indonesia quite a bit, and just to give an illustration — what good central banks do. What they did was to say, "We, the central bank, will fund four thousand universities across Indonesia, to work on Indonesia's key problems." Now that is what I call good QE.[117] That money goes to real people working on real,

---

[117] QE: Quantative Easing is the introduction of new money into the money supply by a central bank.

on-the-ground efforts, rather than say: Let's print three trillion," and make X amount of people rich.

So the issues are complex. And I want to make the point that we also have a mindset issue. As long as we go into this with a mindset that "I am right and you are wrong, and therefore you must follow my order", we will not find the paths to viable solution. So we really need to have a serious discussion on this complexity, interdisciplinary, multi-generational, multi-trans sectoral framework, in which we encourage different ideas to be tried. And then we begin to sort this out as they experiment. If a large company just says, "I'm only going down this way," I do not think it is going to work. But the issues of water, finance are all complex system issues. And if I listened correctly, if you privatise everything it does not work, if you state-own everything it does not work, it is a hybrid. So we are in these blended, hybrid, messy situations, where we just need to let a thousand flowers bloom. We do not know; we should have the humility to acknowledge that we do not have the answers.

> **Question by William Krull**
> *How do we combine interdisciplinary, intergenerational, inter-sectoral approaches to find pathways to viable solutions?*

> *Recently one of the biggest miners and carbon producers in Australia (a family-owned, publicly-listed company) basically made the announcement, "We are going to go 100% hydrogen". When businessmen really begin to understand the need to move, that's probably where part of the solution is going to be.*

## Cycle times, mental models and trust

Individuals have cycles; we have mental cycles. *And our local communities have different cycles that live with our environments.* Come the global collectives, they have very different cycles — they have been driven by the shareholder value cycle, quarterly cycles. The Walmarts and the Amazons of this world are destroying local ecoculture because the cycles cannot cope. That is why we are losing the trust. The cycles of the big collective have not just instrumental goals about money; preserving these collectives is itself a more important goal than actually delivering. If we think of the world as a giant, open, complex system with very different cycles operating at different times, we have now a very growing disconnect of the big collectives — Coca-Cola, World Bank, Central Banks — completely not in sync with what is happening on the ground. That is where the breakdown really comes from. That is where the loss of trust has come from. You promise to deliver equality, freedom, prosperity and all you deliver is disaster. We are really having this cycle of dissynchronization. Instead of the general equilibrium model, we are far from an equilibrium. And we do not have an Einstein to bring all these disparities together. The result is total chaos. And we do not have a mental map about how to get out of this mess. We know it is related to the morals, because *the Cartesian split — mind and body — took humanity out of the equations.* And the minute you take humanity and nature out of the equation, you destroy mother earth, you destroy human beings, you destroy their values. All you care about is money as the bottom line.

The problem is that we thought that science would enable us to figure this one out, and now we have come to realise that science is only 50% of the answer.

There is something besides science, or at least the conventional science that we are talking about. That is the non-science, the religion, the values. And of course, with 7.6 billion people out there, we have

---

**Peter Robertson**

*As Andrew said, it is a mindset issue. In the end, we need the proper mindset, we need trust, this understanding of one another to cooperate. Thoughts about values, trust and mindset are strongly substantiated by Anthony Judge (https://en.wikipedia.org/wiki/Anthony_Judge), ex-director of research at the Union of International Associations (https://uia.org/). Since the late 70s, he analysed thousands of UN projects, wondering why so many failed and how so much money got lost in mismanagement, corruption and greed. He concluded that there is a lack of values. He then looked for a definition of human values, like integrity, respect and trust, and did not find any. Then he hypothesized that human values like trust and mindset are emergent properties, like behaviour, in social ecosystems.*

**Andrew Sheng**

*This is what is so destroying for me. To a large extent, the mental model that we have is not explaining what is happening to all of us. But we need such a model because we are the divided self — we are individualistic but at the same time we are "we". But in the "we", we do not see any trust. We cannot trust our employer, because he could sack us tomorrow; we cannot trust the government, because they do not protect us.*

7.6 billion opinions. The result is this massive complexity. How do we deal with this? I do not have any alternatives, except to go back to the Chinese philosophy of which I am a little bit familiar. It is either you have *youwei*, which means action, or you have no action, *wuwei*. The easiest way to think about *youwei* and *wuwei*, is the Balinese surf. The surfer, when he is on top of that wave, and going through the tunnel, he has no action whatsoever. He just stands there and lets the wave carry him. And it is beautiful, it is poetry in motion. But to get there, he has to strive. And that is where we are. We have a tidal wave of climate change, threat of nuclear war, and all the problems that we have. And yet, certain things we cannot change, and certain things we can. So we do what we can. But we need to understand that into a common frame. That may be the synchronicity[118] that Carl Jung described in the preface to 1924 translation of the *I Ching* by Richard Wilhelm. Synchronicity is the synchronisation of different cultures, different cycles, that produces a new path, whatever that is. Call it incentives to go on the right path. Spaceship earth is hurtling towards disaster, how do we steer it in the right direction. The current mental framework of the captains and the engineers are clearly in conflict. That is where we are, fundamentally.

> **Chinese innovation**
> China is a lot more than its 1.4 billion citizens, it is also a gigantic source of innovation. You can praise Jack Ma for that. He just happened to be there to release the energies of all these nameless, guerrilla innovators who came up with this fantastic stuff that we just go ga-ga about. This could only be done by a collective. It is not done by one director, that is for sure.

> **I Ching**
> The book of Change, nearly 3000 years old, is one of the prominent classics of Chinese literature. Since its origins, its 64 chapters have mainly been used for divination. They still are. The book was introduced to the western world in the early 1700s. Leibnitz was probably the first philosopher who introduced the book to the western world.

---

[118] Synchronicity describes circumstances that appear meaningfully related yet lack a causal connection. Also: the occurrence of meaningful coincidences that seem to have no cause. For more information, see https://en.wikipedia.org/wiki/Synchronicity.

So far, this group has analysed that there is no technological constraint to doing all the things we need to do. And there is no financial constraint to this either. The real constraint is mental. We need to recognise that people are all different. How do we bring them at least *to agree* that we are now living in one earth. That part is really quite critical, finding the common language. It is not just incentives. We are not going to convince people through rationality — that part I have accepted completely. There is no rational argument that would get us on the same page. We need a Leonardo Da Vinci — the guy who integrated art and science — and we do not have that. We do not have in this world anybody who is able to bring the Guernica, that signals to young people — from different cultures — the horrors of war. So we can talk amongst each other; but suddenly realising that it is very difficult to bring us all together. And maybe, that is when — let a thousand flowers bloom. Let different ideas come out, rather than trying to find a one-size-fits-all solution.

# 18   Conversion or alignment

### Cherie Nursalim

When our foundation[119] started, we worked with the Massachusetts Institute of Technology (MIT) on a process called New Theory and Systems Approach, Systems Dynamic, Systems Thinking. The idea is to bring people from different sectors together and take them on a journey. We have worked with various universities such as Tsinghua University in China and the Bandung Institute of Technology in Indonesia, where students must do prototypes together as part of their graduation requirements. Through this, we have seen quite massive transformations in both countries, and across sectors and Sustainable Development Goals (SDGs) — this was pre-SDGs when we started. Later I also became the chair for the United Nations Sustainable

---

[119] The foundation is called, "United in Diversity". For more information, see www.unitedindiversity.org.

Development Solutions Network Southeast Asia, and subsequently we also worked on SDGs, and aligned it with local values.[120]

*Tri Hita Karana, translated as "three causes/ways to happiness or harmony", encompasses the three most important aspects of life in Bali, harmony with God (spiritual), harmony among people and harmony with nature (ecological). Often quoted as a basis for sustainable development.*

*Tri Hita Karana Happiness Chart. 17 SGDs divided into three areas: spiritual, people, ecological.*

SDG PYRAMID

SPIRITUAL

ECOLOGICAL

PEOPLE

SUSTAINABLE DEVELOPMENT GOALS

United in Diversity Creative Campus © Kura Kura Bali
© Copyright 2020

For the Balinese, these values are embedded in the Tri Hita Karana,[121] and aligned with values in other cultures and histories (such as Hindu, Chinese culture, the Pope Laudato Si and even Islam). Our faculty has spoken with Indonesian communities across the archipelago and what we have seen is that *we need to build capacity for the stewards — they are the stewards of the earth. We need to build capacity for our youths too — they are the stewards of our future.* So, we realized there is a lot that needs to be done. In Tsinghua we worked on the Jing Jing Ji[122] to align cities or provinces with different levels of development and think about sustainability issues. In Indonesia we have also made major shifts, including on the political system front, and I think one of the outcomes or realization for us is that there are a few areas that need to change. We call it the Happiness Index or Happiness Impact, to align the three ways to happiness and sustainable development goals. And we say ABCD — how to Align, Blend (multiple stakeholders, blended funds, etc.), and Change the DNA of enterprises, funds, governments. That's what we are working on.

We also need to go to the individual, to the self, *to change the mindset.*

---

[120] For more information about the United Nations Sustainable Development Solutions Network, see https://www.unsdsn.org.

[121] Tri Hita Karana is the traditional philosophy for life on the island of Bali, Indonesia. For more information, see https://volunteerprogramsbali.org/tri-hita-karana-the-balinese-philosophy-of-life/.

[122] A metropolitan region in China.

We started working and creating with the government and we have launched the largest blended finance event ever in the world around World Bank — International Monetary Fund meetings in 2018.

*The OECD definition of Blended finance: the strategic use of development finance for the mobilisation of additional finance towards sustainable development in developing countries.*

For more information about Blended Finance, see https://www.oecd.org/dac/financing-sustainable-development/blended-finance-principles/

There are a lot of funds, but they go to the wrong place, so the idea of blended finance is, how do we bring catalytic funding from philanthropy as well as the multilateral Development Assistance Committee into catalyzing private capital into the right projects. Indonesia launched the SDG Indonesia One SMI One Fund[123] under the Minister of Finance, Sri Mulyani, and they have already raised approximately USD3 billion.[124] For the blended finance we are working with WEF[125], OECD[126], World Bank, all the multilateral development banks, and aligning also the businesses and a few others, they are all on the UN Global Compact.[127] They are all aligning. We really believe that it is happening.

## A real example: protecting the forest

In our program one of our fellows, his name is Bambang, who is a government official in charge of national parks, did a project with

---

[123] SDG Indonesia One is a platform that includes four types of pillars that are tailored to the appetite of donors and investors, namely: Development Facilities, De-Risking Facilities, Financing Facilities, and Equity Fund. The platform aims to raise funding from investors, donors, and philanthropist to be channeled to projects in Indonesia that support the achievement of Sustainable Development Goals.

[124] For more information, see https://www.kemenkeu.go.id/en/publications/news/this-is-how-indonesia-funds-sdgs-creatively/.

[125] World Economic Forum.

[126] Organisation for Economic Co-operation and Development.

[127] The United Nations Global Compact is the world's largest corporate sustainability initiative. For more information, see https://www.unglobal-compact.org/what-is-gc.

other stakeholders, bringing them down to the grassroots level. (We have a cross-stakeholder kind of program — businesses, NGOs, others). They went into the area of an indigenous community. This official used to jail these people because they were cutting down the trees of the national parks. He went there, and for the first time, as a senior official, sat down in the homes of these communities, and looked at them as equals and talked to them. And then he realized that the way they operated was actually protecting the forest. Even though they are cutting down the forest, they had a system, and it is very low impact, it is their way of herding and farming. And they came to be partners; instead of jailing them, they now come together to protect the national parks. And this has led to new laws that have been set up for Indonesia.

## Another example: Solar power

| *Cherie Nursalim: What went right* | *Gary Dirks: What can go wrong* |
| --- | --- |
| We had the experience setting up solar panels, after the 2004 Tsunami. Our foundation, our community, we lived in the houses of those people. So, when we rebuilt, and when we set up the solar panels, it was done in a way that was creating an economic model for them, so that it is in their interest to operate the solar panels, because they can make a living out of selling electricity. So if you do multi-stakeholder approach, and you work with the grassroots, and put your hearts into it by living with them, talking to them, engaging and empathizing with them, then the solution will come out differently. | Many people may not be aware of it, but one of the biggest energy projects in the world is in Papua. We put up solar panels — many solar panels — in nine different villages, and it caused all kinds of problems. To start with, we put them up, but we did not have the capacity to train people to make sure they worked. The first thing that happened — local traders came and stripped all the copper out and took the batteries, and that was the end of the solar panels. We also put in our water systems in these villages and the next thing that happened is, they could not keep the solar panels operating when they needed the water, and that created all kinds of fights in the village about who gets the water and how these things should be done. Then you start fixing some of those problems, and created a social problem because now all of a sudden it was the kids who knew how to keep the solar panels and the water system running, but elders were not-at-all happy about that. |

***Gary Dirks, a propos solar power***: These are really complicated social problems. And it is surprising how few groups of people actually think about it as, "I'm not putting up a solar panel; I'm creating a social movement." And the whole thing has to work, or no part of it will. And that is where I think the more difficult things are coming in, because that plays out over and over and over, not just in indigenous communities. I have got it right here, right now. Do we do distributed solar? Or do I work with people in five square miles and put in four gigawatts of solar? What do we do? Big social question. Without belabouring it, those are the kinds of things I think we now have to think about.

> **Peter Robertson**
> *We typically plan on a high-level and we miss the requisite variety at the bottom. The reality is every self-organising system is decentralised, and whatever we implement on a high level should not neglect the requisite variety we need to have at the bottom of the whole thing.*

***Cherie Nursalim, A propos cycles and trust***: I love what Andrew said, that there are different cycles. Definitely a problem of the world has been how businesses have followed a different path of not caring about the multi-stakeholders, just looking for profits and only for profits or incentives. But I believe there are ways to do both, you can have better business and a better world. But first, the leaders have to have the mindset of aligning nature with humans and with the spiritual. So I think we need that harmony of the leaders, of the business leaders, of the government leaders, of the academic leaders. I think you have to come back to the mindsets, and the trust. Actually the late co-founder of our foundation, Aristides Katoppo used to say that its mission is: "Building trust for our common futures." We live in the commons, which are threatened. So we have to build for the common future. And we have seen many successes in using this methodology, systems approach, youth theory, there is a huge community around the world, in Africa it is called the Ubuntu,[128] in Europe, in Switzerland, it is all over the world, these communities. And they are working from the grassroots level. And it is like what Andrew said; let the flowers bloom.

**Bali under threat**

*Developments like mass tourism, climate change and waste are causing threats to the Balinese way of life and it culture. The world needs to help Bali to save its culture, to keep it intact, because it is so beautiful, and it is so precious. If we solve this together, we can learn a lot from it.*

We do believe this works, so we want to build a whole systems campus, multi-stakeholders approach. And Bali is the right place because I have not seen any other place with that strong kind of harmony of people with nature and the spiritual. So we think the world can learn from Bali. We are launching these types of programmes for businesses, for government leaders, for civil society leaders, so that they will have the hearts, not just the heads, but the hearts, and the hands, to take the actions to make the change together.

**Encourage private and public exposure of opposing vested interests, and actively support the definition of policy and particularly private action to circumvent their opposition effectiveness.**

---

[128] Mugumbate, Jacob and Chereni, Admire. *African Journal of Social Work*, 9(1), 2019. Ubuntu philosophy, forms part of the knowledge and wisdom of how African communities and families raise children. Ubuntu represents the worldviews of indigenous black populations of Sub-Saharan Africa, transmitted from generation to generation through observation, experience, language and art. The widely acknowledged maxims *I am because we are* and *A person is a person through other persons* indicate that relationality is a crucial ingredient for human excellence.

# 19  Incentives

*Alexander Zehnder*

At the end of day, we will not be able to change the humans as they are. They are survivors in a relentless evolution. And we know that humans react to incentives.

I am struggling with the current global situation. Let me use an image that I already used earlier. Humanity drives straight against a wall. We still have time to avoid a crash, but little efforts are made to do so. I strongly believe humanity will be able to escape a full crash. But, the longer we wait the more severe the collateral damage will be. Humanity was in this situation many times during evolution. We made it each time but often at a very high cost of lives. If at one point we would not have made it, we actually would not be here anymore. In the past the crashes were local or even regional, but now the danger is global. In the past, humanity had the chance to evade difficult and hard decisions. People could emigrate, out of Africa, to the Americas, to Australia. Now we are stuck. Space, with other planets, maybe too far down the road. Our global community needs to make a few difficult decisions, postponing them will only increase the potential collateral damage. The question is, how can we build systems, or how can we create incentives helping the communities to start to turn our mindsets in the way we need to do. I am looking for ideas for where we should go to achieve the goals we all want to achieve, in such a way that the happiness, as you say, is at its highest.

*Gary Dirks: Build comprehension*

I think that it is important what Cherie has said, in the way in which she has specifically been working on bringing all segments of society into the conversation. There are good ones and not-good ones, they are all contributing to the fabric of society. And I think that kind of holistic approach that Cherie is talking about illuminates why it is important that we agree about that.

A number of people have talked about breaking things down, involving more people, getting more conversations on these things, and I think what is so important — certainly in my work — is that you can then start to build some comprehension about what exactly have we built, and how, really, does it work. And if we are going to unpick it and create something else, then at least we can start by knowing what is it that we have.

### Alexander Zehnder: The case for water

For most things that we are doing — and I can talk from the water aspect — the technology is known and available. Countries that had good governments also managed their water well. In other words, if you manage the water right, you can also run your country. That is how the ancient civilisations started, and there is an interesting issue. All these civilisations disappeared, except the Chinese one. China started almost over four thousand ago. It still exists as a highly civilized country, while the other civilizations disappeared. We could actually learn from all the failures humanity went through. Why don't we do it? Bali's *subak* systems work great. But what works in Bali, may not work somewhere else. How can we marry the local knowledge with the more global needs? Maybe humans have difficulties to think globally, the evolutionary success factor was up to now the tribe and the region. Changing people's minds is easier said than done.. Better create incentives which guide humanity in the right direction.

### Alexander Zehnder: Including universities in filling the pipelines

It seems that the lack of technology pipelines is hampering progress and makes business as usual more practical and safe. Can we generate incentives for creating innovative pipelines? There are many great funds out there in the world, they are all looking for possible ways of investing their money.

### Andrew Sheng: Academia is incentivised wrongly

Money is obviously part of the incentives. If you do not have the money, people will not do it. But let me give you a simple illustration. Where is the knowledge for SDG? It lies mostly in companies, and at universities. What is the incentive of the university professor today? To get publications in accredited journals, not to work on the field. And how do I get such publications? I have to crunch a lot of data, and get it vetted. I was asked by someone to publish in an accredited journal — I have wasted thousands of hours dotting the i's, crossing the t's. Two years later, this journal paper still has not come out. It is completely outdated, but it does not really matter. That is what academia is now incentivised to do. And so you have thousands of NGOs in Indonesia, in Malaysia, in Cambodia, everywhere, looking desperately for money and ideas. And it is not very much. Cherie will tell you, it is not very much money. Actually it is very little money. How much would it cost for me to fit a solar panel to generate electricity for a distant village in the middle of Indonesia, in Bhutan, in China, India, etc. to generate clean water, internet, to teach the people, etc.? Very little. How much of this money is going on the ground? Peanuts.

### *Alexander Zehnder*

What can universities bring in? Are they stuck because of their internal KPIs to publish first-class papers and less more practice-oriented solutions?

### *Gary Dirks*

I would say we are not stuck, but we are like everyone else transitioning from having won the climate wars to what you do now that you have won. Earlier I mentioned the example of the Navajo Nation where I felt like the university frankly missed an opportunity. To be kind to us — we were asleep at the switch — we had not thought that through.

Now, Andrew's point about publications, we all know, anyone who is in the academy knows, there is an enormous amount of truth to that. And in fact, one of the things I have to be extremely careful about is — I am recruiting faculty members to work on the kind of projects I was describing to you earlier — that I do not jeopardise their tenure case. If they are unabe to publish the work that they are doing, they will not get recruited. Because if you want to be a lifetime academic, and many people do and that is good, you have got to publish, period. You have got to get grants and ideally you have got to get grants as the PI, or the co-PI.[129] And the system is really quite unforgiving, although I would say that there is beginning to be some movement around the world in being more flexible. So what I have tried to do deliberately, with the team that I have, is bifurcate. There is a significant group of people who are not at all interested in being tenured faculty. They are research staff of all descriptions — we of course have to raise money to support them, that has been an ongoing effort — but they see their role as being engaged in these kinds of facilitative processes. Filling the pipeline, making that pipeline work. And in fact, there is another big complex of coal plants that needs to shut down in our state, and one of my team has just negotiated a very complex agreement between local NGOs, the big power company that owns that coal plant or operates it, the Navajo Nation, and now our public utility commission on a package of 150 million dollars that will go to the Navajo Nation to help them make that transition. So that is the direction where we are going, but it is hard work. And it is hard work getting on the inside of these collectives. We probably — two big utilities — probably have fifty people who are inside of those organisations, literally, they go there and they spend time inside them, working on how they come to understand steps they can take that are not life-threatening to these people and organisations.

# 20   Politicization

### Tim Benton

The politicization[130] of issues is really what matters. By politicisation I mean getting "it" as a mainstream political discourse. In the climate negotiation space the combination of Extinction

**Wilhelm Krull**

*At what point is it really useful to enter the arena of politicisation, because as I learned recently, it is of crucial importance to start the interaction very early on, to get stakeholder from various parts of society involved, and then gradually to the representation of democracies. We all know that the democracies have huge difficulties in taking hard decisions. So, what would be your pathway towards finally achieving impact?*

---

[129] Principal Investigator (PI).

[130] The action of causing an activity or event to become political in character, https://www.lexico.com/definition/politicization.

Rebellion in some parts of Europe and the Greta Thunberg *school strikes for climate* has completely changed the dynamics compared to 5 years ago. It is now recognized that as the younger generation grows up and is clamouring more for greater action, then that is going to turn into votes. People, politicians are feeling under pressure. They can see a youth movement, they can see a youth vote, they are often under pressure from their children at university or school getting further leaned on and peer pressured and all the rest of that. For me, it is the issue of getting it into mainstream discourse so that it becomes an issue politicians are competing about when it comes to setting manifestos in the run up to elections. So, we have competitive environmentalism rather than greenwashing which has typically been the situation for the last decade or so.

From my perspective, the big issues of the time, from environmental issue of climate change, biodiversity loss and the land use implications thereof, and also the health epidemic that comes from both covid-like pandemics and non-communicable diseases, all got a root around food.

So getting our food system right, I think is key to getting our environment right, dealing with climate change and dealing with health issues on a global basis. And because of that concatenation, I think there is a lot of scope for recognising that as individuals, as citizens, we typically care about all of these things in a kind of deep way, while the political discourse tends to be siloed, very much so that you have a minister of health that does not know anything about or care about a minster of environment.

For me, part of the politicisation issue is driving the transparency

> **Tim Benton: Covid and complexity**
> Covid may have been quite instrumental in changing attitudes. There is probably an environmental signal behind it, and it locked people in rooms where they have had more time to consider a whole range of things. During the course of the lockdown climate awareness and willingness to deal with climate and biodiversity as an issue have risen and certainly attitudes to our food system may have changed. The food system is complex. Complex systems tend to lock themselves into a configuration that is highly resilient to change. Lockdown and Covid in general have loosened some of that, so we have a window of opportunity through which we can build back better if we can push that slightly open door, and make sure that opportunity is not lost.

so that citizens recognise that health, biodiversity and climate are all interrelated. In the name of our children and grandchildren, or in the name of our insurance bills in the next ten years, we are saying we got to do something about this. We are signalling to politicians that we recognise that this comes with lifestyle changes, but these lifestyle changes might increase our wellbeing; they may or may not decrease our wealth, but they might increase our wellbeing since we are in a better world.

We see in global polls, in European polls, in country polls that attitudes are changing. How far these changes go, I do not know, but I think there is a unique opportunity now to get the arguments mainstream. For me, politicisation is about mainstreaming those conversations and making sure that people join the dots. It is not getting everybody to block the roundabouts and stop people driving and climate activism.

> **Accept that fairness should not only be perceived, but also proven for current and future generations through all policies and on the ground action. This should be the essential basis to build public trust and reach the tipping points to modify public attitudes.**

### *Jan Staman*

Yesterday (Chapter 4) you gave a great insight in all the negative feedback loops with regard to food. And you painted for us a lot of misery. So, my question is, what is necessary to get something like this right? Because the way you are phrasing it now, how do we get our food system right? That is a question I think everyone would say yes, that is a good question. But when you talk about these negative feedbacks loops, everyone is going to say, and certainly the farmers, **stop here**, this is not my business. We want to talk about getting the system right. So there is a kind of keeping out the negative things, and putting in a positive narrative. So how to get mainstream? How to get to the point that society is talking and can accept that these are the problems, and that the only way to get out is giving a new answer to this question of yours, "how do we get it right?" And what is necessary to keep it mainstream, not polarised, just like the well-organised democratic debates we usually have?

**Tim Benton**

Part of the answer is about developing a common vision. One of my criticisms of the UN and the sustainable development goals is that they hand-wavy talk about the world we want, but no one ever defines the world we want, and no one recognises that there are trade-offs between the 17 goals. I think there should have been an envisioning exercise to say, what is positive about the future. Even within the climate discourse narrative, you still have people saying that the future is going to cost, it is going to be painful, you are not going to be allowed to fly, it is going to be dreadful. But the positiveness of the future, that we will actually suffer far less mortality because of air pollution, that we will not have the climate anxiety, etc., has never been well articulated, because it is articulated in different places by different special interest groups.

> **Let multi-background teams, supported by scientists, translate existing global climate change projections into transparent (multiple) national and sector long-term scenarios to guide future progress on addressing climate change largely independent of existing bureaucracies.**

The other part of the answer is recognising that, over a 70-year period, the current system has been designed deliberatively to drive consumption growth. It has been designed that way, so it can be undesigned or redesigned. It is not a God-given thing that we grow lots of calories, it is a subsidy given thing that we grow lots of calories

> **Transition to the end of slavery**
> *To abolish slavery, the UK paid every slaveowner 500 pounds for each slave (£20 million pounds in total). It was not the aristocracy that had the slaves. It was pretty much people in country towns that had 1 or 2 slaves as investments in the plantation work overseas. Giving money to slave owners, to get them to give up owning slaves fuelled the industrial revolution in Britain, because it provided a whole lot of ready money for people to invest in other things outside of slaves.*

and it is an unlevel playing field. So, it could be changed. But it is a complex system, so it is locked in. And there is a huge number of vested interests, because of the economic leverage that comes from the consolidation of companies into bigger and bigger entities. The big vested interests (collectives) have incomes greater than half the countries in the world. So, we have to start talking about just transitions.

**Alternative proteins?**

*Do we need to grow alternative proteins? We do not need to because we have got enough protein in the world to feed everyone healthily.*

*But in a sense, alternative proteins is an economic way out, because it gives adventure capitalists something else to make money out of if we are saying do not make money off chopping rainforests.*

If you think through where we are today with industry, we are proposing something that is going to hurt people and is going to change vested interests. And we have got to find a way to make them stop fighting against it. And that is not giving them a free pass that they can run everything forever.

So, the just transitions, the resilience issue, we need to unlock some of the lock-ins to make the system move, and the visioning and all of that rests on whether people want to do it. That is why I think the political will, the market will come if people are saying that we want these changes. If people are not willing to accept Teslas instead of internal combustion engines, it will not happen.

# 21    Useful models in fishery, conditions for change

**Wilhelm Krull**

*Gert gave us many different examples on how the fishery issue was dealt with (Chapter 5). I almost felt that you were close to a typology. What would be my interests is of course, what in your view is really effective, or is there a catalogue of different options we have in regards to solving transnational issues.*

*Gert van Santen*

The examples that I have given, particularly Japan (Chapter 5), are something that I do not believe you can take wholesale and set it up somewhere else. It requires a culture; it requires a history. What I wanted to show is that it makes sense to break up a complex "fishery" into smaller units and try to work at the local level to improve some of these things. And most of the time, fishermen are quite willing to make changes if they clearly see what the situation is supposed to be in the not too distant future. And also if they see that if things go wrong for them, that they have opportunities and support to make the actual changes.

So, if you want changes to happen, you have to show what is needed in that not-too-long-term future, and what kind of money you need to make those changes. Investments in fishery are heavy investments, and you deal with heavy interests, of all kinds of people. And there is illicit fishing which is a separate issue altogether.

If you just invested in a 15 million dollar trawler, you do not want to change the way quotas are being distributed, because you are totally dependent on those quotas. Unless there is financial assistance to get you out of that loan and out of the business, you are going to continue fishing.

Also, if you look at small scale fisheries, you have far too many people involved in local fisheries. So if you want to change, you have to make proper arrangements, or find different opportunities for the

> **Gert van Santen: An issue of pricing**
> *Looking at land-based protein production like cattle rearing, one of the key problems that exists in the Netherlands is, if you want to have a circular milk production that is based not on producing more and more, and trying to export as much milk powder to the rest of the world as possible, you need to increase the milk price to the farmers, which is apparently being controlled by a few very large milk producers and supermarkets. So how is it possible to make adjustments in a country like the Netherlands to do that?*

people involved. And those arrangements must be such that the people involved can see, okay, it is going to look like this 5 or 10 years from now, and this is the kind of help we are willing to give. I think that is important.

But Japan shows that it is possible to modify the system once there is an external factor that makes the continuation of the present system impossible. And in Japan that is very simple, the average age of a fisherman right now is more than 70 years old. So they cannot man all these vessels. They cannot continue fishing on the high seas for tuna. There is just nobody who is willing to do that kind of work anymore. So they realised: "We have to change it", and that is the other point, in Japan the fisheries agency is very strong, it is well established, it has a lot of power to make changes, and that is the reason why they could make the changes that were needed, and I think they will continue to do that. So those are the two things I would like to say: move most of the distractions to a lower level, but also create at the top possibilities to make substantial assistance to facilitate those changes. And one of them, as Tim also said, is transparency, you have to be transparent about what you are really doing and why.

### A practical approach

*In West Africa are a bunch of old Chinese trawlers that are continuing to fish. The reason is that they do not have the money to sail back all the way to Shanghai! And for that reason, people say ok, you can continue fishing, have some financial income, as long as you drop the catch here locally.*

If you go to West Africa, keeping industrial fishing out of a substantial coastal area, is absolutely critical for the supply of fish locally, but so is the reduction of imports of frozen fish from somewhere else. Right now, one of the craziest things is that fish that is caught in West Africa by industrial fishing vessels is moved to Oman, where a bunch of traders basically re-export the fish back to West Africa. Now if there is anything loonier than that, please go ahead. So the question is, how can you change very practical things in the system? I think for such practical matters pressure needs to be applied both from the top and from the bottom, whether its milk prices or the exporting/importing of illicitly caught fish.

### Sander van der Leeuw: Where to start?

I entirely agree. What I would do first is to find out, apart from price, but in people's minds, what is the importance of the fish? And can I begin to find parts of the arguments that are sufficiently open to change? I will give you a very old example, a story from an Indian colleague about how a particular fishery's community began to change the situation. They got into trouble because they were overfishing. At the time the government was very much affiliated with the Soviet Union, and had created huge boats. So what did the villagers do? They created 3 months of religious holidays, during which you could not fish.

We are trying to do some work in Africa trying to begin to get a sense of how to get communities themselves to identify the points where their narratives are changeable.

**Sander van der Leeuw**

*This example, although kind of particular, actually gets us out of the economic and progress thinking, that I think is the real problem with the sustainable development goals. These goals have been set by people from many different countries, but were all educated at the same universities. Their basic idea of progress is not actually relevant for a lot of those communities.*

### Gert van Santen

Many small-scale fishermen in West Africa are hostage to their own income. They need to catch a few fish to be able to survive

**Sander van der Leeuw**

*All these local communities are caught in a global system that keeps them fixed in that trap.*

until tomorrow. Those are the kind of things that really need to be addressed, if you want to make change. The risks in lower-level income countries is a huge issue that needs to be addressed. Unless you can deal with those risks in a believable and practical way, you cannot really change the system.

# 22 Priorities for change, who/what is going to change the system?

### Mark Wilson: The importance of an intersectoral approach

What Gert said points to an important issue we should discuss. We talk a lot about change in systems that we see as more sophisticated and are driven, hopefully, by desire for long term sustainability. We hope that is the case. But, if one compares the UK's ability to impact climate change versus India, what should our focus be? Should it be on industrialised countries or should it be on low-income countries?

I am less of the policy person. I am more of the engineer, how do you implement policy? So I have spent a lot of time at the community level. And where we have success in low-income countries is when you get the communities to be participants in the decision-making process. Setting their own priorities. And there, we can see very quickly that they are very risk adverse. Low-income people are extremely risk adverse, as we all know. So to get them to bring about change, we need to find ways to ensure that they get access to finance that allows them to make rational decisions. Right at the moment there is no incentive for them to stop burning wood, to change from producing rice or wheat, simply because less risk is involved in producing those basic crops. How do we get them to move forward, raising value added to those commodities? I hope that in our discussions we do not get lost in the clouds. We recognise differences between silos, but the people that we **really** need to change, other than the industrialist in the high income countries, are the peoples who produce the product, and have very little incentive to change their production cycle.

> **Mark Wilson: Complex intersectoral discussions or something more basic?**
>
> *What would motivate sub-Saharan Africa to unleash its tremendous potential for the future world food supply?*
>
> *Are these very sophisticated, complex intersectoral discussions that we are having, or is it something more basic? And this comes to Gert's point, where do we start? At the local level or at the policy level? We have had generations of experience at the policy level. Some of that works, some of it does not. But in low-income countries, the weakest bureaucrats tend to be in the ministries of agriculture. They tend to be very feudal in the way they look at things. When we come to them with these kinds of policy issues, they are not the leading lights. Instead they tend to be the ministries of finance, or the ministries of economy, not the technical line ministries, where we really want to bring about the change. **Let us not lose track of the worldwide difference between these two**.*

So, I just want to make sure we do not lose sight of that. The multisectoral approach is extremely important, extremely important. How one will get them on board is the challenge. As I was listening to Gert talking about Japan, absolutely, but I was more interested in his comment on West Africa and how they were focusing there.

### *Tim Benton: Who makes the decisions about the economic system?*

I think the meta question which Andrew effectively raised yesterday (Chapter 8) is: who makes the decisions about the economic incentives, that drive the behaviour of rational people at the levels that make those decisions? From my perspective, which probably matches Andrew's perspective, is that we have got a system incentivising things completely wrong.

Who is going to change this system? It is not the small low income southern countries. We have driven the industrialisation of agriculture from a northern imperialist perspective, the Washington consensus and, I guess, the World Bank as well. We made sure that we will only lend them money if they play our game and open their markets to us, knowing that we can outcompete them, and drive the economics in a way that benefits us first and benefit them second. And the issue is how do we move away from a consumption-based economy that drives these negatives all over the world, towards something that is more based on improving people's wellbeing, of which wealth is a part but not the only part. And so, given the economic incentives to externalise cost and grow the wrong things, in a rational way, to maximise economic returns, how do we change that?

> **Tim Benton: A systemic shift**
> *We need involvement in trade, making the right sorts of investments, **us not being exploitative,** but learning on the ground, building resilience so when things go wrong, you are not in a situation where they are jugged and go into a food insecure situation, and we benefit and just swap our supply chains from West Africa to Southern Africa or whatever it might be.*

I think the problem with the bottom up is that we can always get local examples of what works and we can tweak the system, but that is not the radical shift in the system that we need to get out of the vicious circles. That is local kind of sticking plaster, and from my perspective, what we need is a systemic shift, coupled with exactly what you are saying, local decision making.

So I think we are both right, but also both wrong. Because it requires everything to change, and it is the politicisation of the debate. Five years ago, this kind of thing was not really on the radar. Now there is a fair amount of discussion, partly prompted by some of the climate change stuff, about whether GDP,[131] as a measure of economic flow, is the best way to structure an economy around, or whether we need to do things in different ways. And the Dasgupta[132] report, which was published in the UK recently, says that GDP does not allow us to have a sustainable economic model, because it is just tracking the flows, not tracking the asset base that is being eroded to create the flows. So, I think there is just so much that we can do differently but your bigger point that we must listen to communities in the global south is a well-made one and a well understood one.

# 23    How people make decisions — role of experts

*Sander van der Leeuw*

Over the past few years, I have become more and more interested in how people make decisions, at the level of local populations as much as at the level of decision makers. We know very little about that process. To change decisions, we need to see how biased people are in looking at the complex system that they are dealing with. And those biases are not something that we know a lot about. We know a lot about the end result, the combination of biases, but we know very little about the individual biases, and therefore we know very little about how we can actually change this situation.

> **Sander van der Leeuw**
> One of the interesting things I have been involved with in China, a few years ago, is an attempt to actually get the population in a village to decide its own future. And then for us to simply help them realise that. Of course, the current situation in China makes that extremely difficult, so we could not bring that to an end. But nevertheless.

---

[131]  Gross Domestic Product (GDP) is the total monetary or market value of all the finished goods and services produced within a country's borders in a specific time period, https://www.investopedia.com/terms/g/gdp.asp.

[132]  The economics of biodiversity: The Dasgupta review (2021), Final report. https://www.gov.uk/government/publications/final-report-the-economics-of-biodiversity-the-dasgupta-review.

Another element in this is expertise and the role of experts. At ASU[133] we spent 10 years, trying to convince the managers of the Arizona water system to actually have a look outside their immediate objectives. These managers were so engrossed in their particular narrative, that it was virtually impossible to get them out. In the end we did by paying personnel in the city and the state government to help this along. For me this whole issue relates to what I think is another, much more long-term issue: where is the whole education issue? We educate children from the very first, in a sense of "there are truths", rather than educating them in the sense that "there are always alternatives". The reasons for that have to do with the practicalities of education and kindergarten and things like that. But the result is a really important one, because people are socialized around particular visions of the world around them. And they are told that that is the truth, and they are not told that they could try and find another way.

**Tim Benton**

*I have often had conversations around the former US president. What sorts of evidence would he need to make different decisions? The answer is, it is not an evidential based decision that he has been making at all, it has been a set of criteria that do not reflect what many of us as academics and liberal thinkers would say is the rational basis for making decisions. But it is rational from his perspective I guess, because he is protecting his interests, even if it is in the interest of today, versus the interests of his grand or great-grandchildren in years to come.*

One of the things we have pitched in the last 2 or 3 years, is the way in which in the West we create categories of which some are open to some change, and some are absolutely not open to any change. That dynamic directly impacts on how narratives are constructed. If you look at Shiller's recent book, about narrative economics,[134] you begin to see to what extent do narratives created by concatenation of those kinds of categories actually impact the economic decisions making.

---

[133] Arizona State University.

[134] Shiller, Robert J. (2019), *Narrative Economics: How Stories Go Viral and Drive Major Economic Events*.

**Tim Benton**

So how do we change the economic decision making? Do we change the narratives, or do we change the biases of the people who are making the decisions?

**Sander van der Leeuw**

I think we do both, we are going to begin by understanding what the narratives and biases are. Identifying which parts of narratives are actually open enough to be changed, and which parts of those narratives are not. And then trying to insert new elements in an existing narrative that redirects people.

> **Expand education and research capacity to develop human capital in private and public sectors to redefine the prevailing economic narrative to implement climate related civic, public and private/ corporate policies, regulations and actions.**

# 24　Unblocking blockages, levers and priorities, a limited inventory

**Tim Benton**

What would be the things that would unblock the blockages? Is it financial incentives to do things in different ways, is it changing business models, is it a deregulation or regulation issue, is it changing people's attitudes, is it transparency? What are the things on top of people's minds that will actually drive things in the right direction? Or is it a little bit of everything? What would we prioritise in terms of its ability to move the system in the direction that we want to move it.

### Wilhelm Krull: Identify the drivers of change

As we all know these are complex issues. What I consider as important is that we very specifically identify in this kind of multilevel complex system or complex constellation, what are the drivers of change?

Whom can we rely on in this process of actually making change, and who are actually willing to move along this kind of path towards viable solutions? And my experience is that it depends very much on the circumstances people are in, on their beliefs, on their attitudes.

That is why I think with respect to identifying agents of change, we cannot just say that it is individuals, or institutions or parties, but it depends very much on what the specific circumstances are. The examples that Gert and others gave, show that a lot of it is culturally dependent, dependent on the strengths and weaknesses of the institutions. The real challenge is to find partners or to forge partnerships that could then work their way through the issue. And that of course is something, the newly-to-be-established NEW INSTITUTE, will take on. We will deal with it in a fundamental way, with respect to what are the values of a sustainable society as well as a sustainable economic system, and what is needed in our modern democracies to change in order to return to deliberative democracies, encourage participation, and in particular, also solve the problems of the representation democracies. As you all know, there is a lot of distrust in the functionalities of our parliaments. So, it is a really wide array of different things, but you need to start somewhere. And with regard to political parties, one has to find out to what extent political leaders are willing to take their constituencies with them on a path towards a more viable and sustainable future. And there, fairly often, the short-termism of our electoral systems hinders politicians to take a long-term view. And so, I think, academia has a role to play, but academia in itself or on its own is a very weak transmitter of these kinds of messages. So to forge these kinds of partnerships or to identify these agents of change is at least something which I am struggling with and where I would be glad to learn more on how to go about that.

> **THE NEW INSTITUTE** in Hamburg, is an institute of Advanced Study and a platform for change. Its mission: to imagine and develop visions for fundamentally reconfigured societies. It seeks to close the gap between insight and action, by bringing together academics and practitioners from different disciplines, united in the quest to analyse what needs acting upon, and to act upon analysis.
>
> For more information, see https://thenew.institute/en/why.

### Mark Wilson: Agents of change and/or respected individuals

If we look at Greta Thunberg, she was able very quickly to mobilise a movement among the younger generation, and that is very reassuring. In the United States we see a few individuals, like Bill Gates, who have access to politicians, which is important. Gates is judged by his success but he has moved the paradigm. I sometimes feel that it is not so many agents of change that we need, but having respected individuals in society who are convinced that change is necessary, and who can articulate that to those political leaders in a way that they listen to them for whatever their motivations are. I see that in people like Bill Gates. I wonder if we need to do a better job of identifying these individuals and educating them in a such a way that they see this balance, this complexity that we are talking about, and they then discuss this with our political leaders.

It is clear that there is an anti-establishment approach to politics. Anything that is said by central government, people viscerally react to. Whether it is rational or not, they react to it, that is very true in the United States, that is for sure. And therefore, I tend to look to a few individuals, who have that access, and I gave Gates as an example. Just to build on your last comment there Wilhelm, I think that was a very astute comment.

### Tim Benton: Can we predict other agents of change?

Do we need systems or models, or do we need to examine every single context? Can we come up with any sort of generalities or identified patterns. Anything? I know it is a blunt instrument, but the financial economic power, the way power scales with financial returns, is a key thing in terms of identifying where the blockages are.

> **Influential people and personal interests**
> Tim Benton on Prince Charles: I work a lot with Prince Charles. He thinks of himself as an influential agent of change, but I know that no one listens to him unless it is in their personal interest to listen to him.

On the one hand, influential people can influence, but what are the agents of change, what are the things that might be important, to respond to Wilhelm's comment. But also, where else can we drive the change from. Who else has got a list of interventions that they would like to see, or are they thinking systemic change? Is making the system more transparent, allowing people to make better decisions, is that necessary but not sufficient. Is it about the political power, is it about the just transition? Is it about having a wellbeing economy, and recognising that infinite consumption growth on a finite resource base is impossible? Where are the mindset changes, where are the issues we should be pushing?

### Tony Mayer: Localism

I was just mulling over something about localism. You know, because we got to get communities to buy into this, and it starts at the local level and builds up. But just thinking of the experience where I lived in Swindon in the UK. For those of you that do not know it, it is an industrial town in the middle of an agricultural area. So it has an odd relationship with its immediate hinterland. It is growing rapidly in terms of new housing. During lockdown I do my walks around the place, walking through these new developments. Not one, and we are talking of probably thousands of houses, not one is built with solar panels. Not one is built with heat exchangers. This is a marginal cost when you are building a house, but of course, the developers would say it is pulling prices up.

So it is only at the local level that we can start to put pressure to do the right things there, but if we look at that total housing development, the carbon footprint is enormous, in terms of the infrastructure, the roads, the bricks that are made to actually build the houses. We have a housing shortage in the UK, and we have to address it, but at the same time we are still building up problems for the future, the things that are not sustainable. It is how do we mobilise people at all these levels so that in the end government would be forced to intervene. As Wilhelm mentioned "democracies have huge difficulties in taking hard decisions", so developing local pressure may have some effect. Demographic change will cause change to happen, but it is far too slow. I thought Covid would see a dramatic reduction,

**Tim Benton**: *Listening to discussions on the radio, it seems to me that the predominant narrative right now is that we are better off the more money we have, the more we spend, the more goods we purchase.*

but that has not occurred. The impact of Covid lockdowns worldwide on carbon emissions has been marginally only.[135]

### *Wilhelm Krull: Back to normal?*

Covid, the pandemic, is an opportunity but at the same time, back to normal is kind of the catchword of the day. What I would be interested in, is how to keep the momentum of the awareness that we cannot go on as we did before, and not lose this impetus as we did after the financial crisis of 2008/9 when the financial markets returned to their traditional ways of dealing with things. For me, this is the kind of tipping point we should address.

### *Tim Benton: The rules of trade*

With regard to financial markets, certainly in the area of food, the lowest common denominator that drives the system, other than our general economy, are the rules of trade. These rules of trade have taken decades to get to where they are, and they are not fit for purpose in the modern world. It will take a long time to deconstruct something that has taken decades to put together. So I despair sometimes that we will not get out of this kind of lock-in[136], until something serious happens, and obviously Covid is not serious enough. Lots of people have said we must build back better, but almost no one is building back better, and almost no one is investing in building back better despite the rhetoric.

### *Gert van Santen: On Chatham House*

Tim, as an institution, Chatham House, which is basically a technical assistance institution, do you think it can carry or influence things at the levels that are necessary? At the top but also at the lower levels. I mean do you have the money to do that or do you have the connections and partners to do that?

---

[135] Tollefson, Jeff. *COVID curbed carbon emissions in 2020 — but not by much.* Nature 589, 343 (2021), https://doi.org/10.1038/d41586-021-00090-3.

[136] Lock-in refers to a situation where it is not possible to end or change a (financial) arrangement, https://dictionary.cambridge.org/dictionary/english/lock-in.

### Tim Benton: Chatham House

At Chatham House, every day, somebody from our UK government or another government rings up and asks for advice. Even when I worked for the UK government, I could not get the sort of convening I now have, because we are seen as a trusted institution or a trusted friend or facilitator of discussions. Very often when people are scratching their head about how to tackle a problem, they will reach out to a place like Chatham House, whereas they would not have reached out to me as an academic, because they always say academics have vested interests, in the sense that they (academics) will say, "No I don't know the answer but give me a research grant and I'll be able to find out for you", and then come back three years later still not having the answer but saying that "I've had a very interesting time, but give me another three years' money and I'll answer the next question."

So I think, with the level of influence Chatham House has, you cannot change the system from within, but you can at least influence it, and nudge it in the direction you need to go.

> **Agents of change: Chatham House, World Bank, Universities?**
>
> *Chatham House is seen as a trusted institution or a trusted friend or facilitator of discussions. With its level of influence you can't change the system from within, but you can at least influence the system, and nudge it in the direction it needs to go.*
>
> *The World Bank facilitates changes in the way things are operating in the field, providing people that can make change with the tools to do that.*
>
> *Academia has a role to play in being an agent of change, but, with some noticable exceptions, universities tend to move away from reality in order to be theoretically correct, and thus are weak transmitters of the message that change is necessary.*

### Gert van Santen: The World Bank as an agent of change

There is a similarity with the World Bank. One of the things I really enjoyed as a World Bank expert working for a specific project, was being in the field and make changes to the ways things were operating, at almost an *ad-hoc* basis. I think when you talk about systemic change, making relatively quick decisions and providing people that can make change with the tools to do that, is one of the things that can make a change.

# 25    If you could, what single thing would you change?

**Mark Wilson: I would make all these different silos sit together and formulate policy.**

Just listen to each other, and at that level formulate a strategy together. And then from that you build your budget, and allocate resources. If I could do one thing as a president of a country, I would do that.

**Gert van Santen: I would like to know who my representative on earth would be.**

I worked with a lot of "ministers" in a lot of countries, and there are huge differences in their ability to make changes. Some of them are very good at it. Many others are not for one reason or another. We need a system whereby we can identify people that do have the ability to change. And whether they write books or are high school kids from Sweden, there should be a system whereby these people can be encouraged to continue and have some tools to continue. And I am not at all sure what kind of agency or what kind of process should be able to make that selection. But I think that unless we are able to pick out those agents of change, we are not going to solve this complexity problem. We have to have a limited number of people that do have the kind of connections and discussions to determine what items can be changed, and change those.

Even at a very low level, if you go to cooperatives in India, you will see that of those cooperatives on the coast, about half of them are very sloppily run and not very good, and there are about twenty percent that are very well run and can make a difference. It is those twenty percent, and the guys that are in charges of those cooperatives, that need help in order to make a change and get a vision for the future. I think there should be more agencies or clubs, that are able to determine who are agents of change.

### *Jan Staman:* I would change the narrative

Tim said (Chapter 4) that we need a philosophy on what makes agriculture right. And Sander asked how do we get that discourse in the mainstream of our countries, of our public and political debates. Sander also said that one requisite for that is that you know which narratives are allowed and open, and which narratives are closed, and that you have a big problem when you do not know. So the first thing you got to explore is what kind of narratives are in a change mode now. Maybe you cannot say they are open or closed, but what kind of narratives have promising traits with regard to what we are dealing with, and can we get directed to those kind of narratives? That to me, seems very important for politicians, because that is the only thing they are good at, finding these narratives and trying to get connected to them. And, whether as a scientist, or a group, or an influencer, if you can offer that and show that you can help them explore in this way, then you are are getting into the mainstream of the public discourse in your country.

What you need for that is a sense of urgency. This sense of urgency is there, at least in the Netherlands. Agriculture has become a wicked problem[137] and very often governments need years to solve minor issues. But in the Netherlands, in the case of nitrogen deposits, the judge ordered the government to stop with these slow-down practices. And now all the old narratives are completely closed, and everyone is in need of better narratives, narratives stressing the urgency of change. They are all looking for sources for such narratives. Every farmer in the Netherlands knows, and is saying so in the newspapers, we got to change, please help us change. And then you have that

---

[137] The term wicked problem originates from social planning. A wicked problem can be loosely defined as *"A problem where the problem definition and the solution cannot be kept apart"*. The distinction between complex problems and wicked problems is not very clear. One way of looking at it is to say that wicked problems are a set of special problems (related to the functioning and planning of society) within the more general set of complex problems. See also: Horst Rittel and Melvin M. Webber, *Dilemmas in a general theory of planning*. Policy Sciences, 1973; 4:155–169.

big agro-industry which cannot provide hope yet. And the Rabobank, one of the leading agriculture banks, is saying, we want to help you but we cannot. (The farmers visit the minister and ask: "Please help the Rabobank"). This the situation we are in now.

New consortia of the willing are coming forward. There is help from Brussels. New storylines are tested. In new consortia and in public and in the political debate. I think we will be very successful when we find the proper narratives. I think that is how we can get agriculture right, how we can take care of not losing too many farmers, of getting more farmers who are earning more money than they do now. Small farmers with more money, so to say. Farmers with fewer animals and more money.

How are we going to organise this and how are we going to create an agriculture of "needs", instead of agriculture for the greedy? How do we reconnect agriculture and health? How do we get farmers who are proud of solving our health problems, which are enormous? I would say, go to the stakeholders.

I am sure, when you are going to the Rabobanks now, and the greenbanks, and to some big industries too, and sit together and say: "We need to get agriculture right, how are we going to do it, we being Germany, France, Belgium, the Netherlands, Sweden and Denmark, I would say the problem is solved. Yes, we are in need of a paradigm shift.

### Tim Benton: It is about the narrative and the vision

I was an author on IPCC special report on climate and land, and what really struck me in the plenary session was that almost every government delegation had a representative from their ministry or department of agriculture, but when you talked to them, most of them described agriculture as a trade ministry, whose job was to use the land as an outdoor factory, to produce the commodity, to export for economic growth purposes.

I think that, certainly in my government, a very dominant narrative is about growing more and more and more. Sometimes that is twisted into "we got to feed the world" narratives, and that is a very different narrative or vision for the future from "food is supposed to be about nourishing", or "we are guardians of the countryside", which is another one of those farming tropes that are quite powerful. You are right, it is about the narrative, Sander's right as well, it is about the vision. But it is also about removing the disincentive for change. Because the amount of times I have had farmers come to me and say "my grandfather would be turning in his grave at what I have to do to make a living. I would love to be doing things a different way but I can't make the finances work." So, somewhere in between changing the business models and the incentives, and the narratives, lies the key to changing agriculture.

### Jan Staman

That is all true. But Dutch young farmers want change and they want autonomy in order to take responsibility for that. But in the newspapers and in public debate you will for a rather long period of time still find the parts of the old narratives and self-descriptions. When you are going into secrecy, in silence, and organise the willing, you would be surprised, in the Netherlands at least, because the atmosphere is a highly urgent feeling that we have got to change.

### Tony Mayer: I would be a (good) dictator.

Habits have changed, expectations have changed, farmers still got to make a living, and yet the whole thing is out of balance. Somehow, saying what would I do, I think I might just throw my hands up in despair, because it is incredibly difficult to square that particular circle. You want to improve nutrition, you want to ensure people are prosperous enough to have the food that is nutritious. At the moment the whole system is that non-nutritious foods are the cheapest. So the populations of the poor eat poorly, and then they suffer from obesity and other diseases. Somehow we have to break into that circle.

> **Tony Mayer: The unbalance**
> *Britain is seeing a rise in people needing to get basic foods through food banks and charity. Yet, I am impressed by the cheapness of the food in the UK. Most of it must actually be on supermarket shelves at a loss. For example, you can go in and buy a quite substantial chicken for 3 pounds, 4 pounds, which is approximately 5 euros. Decades ago, you had a chicken for Christmas, not every week.*

**Sander van der Leeuw: I would go for the advertising industry.**

I think the way we are semiconsciously or unconsciously influenced by the fact that big corporations have the money to do the advertising. This means that we have lost a huge possibility to impact the situation. And that is aggravated by the information technology revolution. The fact that everyone now can express their own "truth", has made for an incredibly more fractious society which, I think, ultimately will come to some sense of reorganisation. But I think we are in for several decades of a really huge mess. If we are really able to actually deal with that, and re-establish what we had in the west until let us say the 70s — that is a degree of control over opinions by newspapers and by televisions, that was limited, and you did not have 300 TV channels to escape to if you did not like what was being said — I think we would be closer to being able to get people to think about this in a different way. So that is where I would start.

**Tim Benton: I would fundamentally deconstruct the system and start again**

I would toss up questions like build the economy around wellbeing, internalise the externalities, in terms of regulating economic costs of unsustainability and poor health outcomes into our food system, or maximise transparency so the people can make the market work better. In a sense, you are controlling the information flow, and targeting marketing. I think it is part of the issue that we have got to recognise ultimately, that you cannot have exponential consumption growth on a finite resource base without something breaking. So I am curious as to why most of your suggestions are somewhat incremental rather than transformational, in the sense of staying within the system, my inclination, would be to fundamentally deconstruct the system and start again.

**Jan Staman**

I agree with you, that is what I hope, transition, and deconstruction. Now deconstruction is accompanied by losers and winners. And you need political mainstream debate to organise a perspective for the losers and the winners. How to give hope to losers. And this is what every politician knows: when I am not going to do that, they will kill me. So how to do this? And I would say at the very beginning it should be a language of hope, not of deconstruction and killing, but of hope. And during that process, credits will grow for transition and deconstruction because the mindsets of the people involved can accustom to the state-of-the-art.

# 26 Do we need a real crisis?

### *Gert van Santen*

There was a famous Dutch football player by the name of Johan Cruyff, who had these one-liners. One of his famous one-liners is: Every negative has its positives. And I think that, as far as food is concerned, we are not at the crisis level yet, in the sense that there is still enough food, and the people that do not have enough food do not have a say in anything, so we do not have to really worry about them. A real crisis in terms of

**One liners from Johan Cruyff**
- *Often something has to happen, before something happens.*
- *You only start to see it when you understand it.*
- *Truth is never exactly as you think it is.*

food supply, like in the Middle Ages, when there was not enough grain and everybody was hungry, that can make real change. But I am afraid were not there yet. Even in West Africa, the small-scale fishermen, they still earn a little bit of money. It is practically nothing but they are able to continue on that basis. And sometimes they do not have any money, but that is not sufficient to take over the ministry of fisheries or agriculture. It may get there in the not too distant future, but I do not see it happening yet. And in the few cases where they tried, they were simply moved away and nothing happened. I think that it is also a concern that, if we really want to have change, we need to have a real substantial crisis, for a lot of people, not only the poor people but also the rich people, and I do not see that happening yet.

### Tim Benton

So one suggestion is that we pray for war, or something like that in terms of a global crisis big enough to deconstruct the system, and then we come out of it with a greater incentive to build back better.

### Gert van Santen

The 2nd World War actually did that in the Netherlands and in Europe, no doubt about it. It is terrible to suggest that, but it is one of the things that came out well, after the war.

### Tim Benton

As a systems person I would like to deconstruct the system and build it back better, but we cannot really wish for a global crisis. You know a few years ago I would have said that something like COVID-19 would have been an opportunity to change direction. But it seems to be that no one is taking advantage of that opportunity. People are saying they are but not in reality.

### Jan Staman: Deconstruction happens all the time

Deconstruction happens over and again. Look at the oil industry that is going to disappear. And the gas industry in the Netherlands, it is over and out. It is just happening in a few years. And it is real deconstruction that is made possible at the very moment, that this call for change is getting a mainstream character.

So, we do not need war or disaster. There should be something like "What else can we do about this?" And then we deconstruct and start again.

### Tim Benton

We are doing a lot of work on energy transitions and one can counter your comment about oil and gas by saying that China and India have coal fired power stations being commissioned left right and centre for the next ten years, Venezuela is opening up new oil fields, Norway is going to drill in the Artic, and what is going to happen to American oil reserves? We might have moved on a bit in Europe but that has not dealt with the climate crisis. And likewise tobacco, Zambia is still paying farmers to produce tobacco for export because there is a global market for it, even if in the rich world we have largely moved away from it, but it moved over to being rife in Asia and other parts of the developing world. So locally, industry comes and goes, it is more difficult to find an overall example of where something that has radically changed, other than the anecdote I have made earlier about slavery. That was a global issue, but it was largely driven by some very large colonial powers, so where there was a kind of obvious global governance change.

### Sander van der Leeuw

I would suggest that there is a way to go about this differently: Correct the imbalance between speculative finance and real investment in the right kind of production, to get the kind of wellbeing society that we are all talking about. I would argue that that would take a relatively limited legal adaptation, that it would not necessarily make huge loses, if it is phased in sufficiently coolly. I personally feel that it is the financial industries where we need to attack first, because they hold the keys, they are the most shameless, and they have very often no idea about a wider vision of what is happening in the world. And so for me, in our western system, that is a weak point which we can and should be attacking.

> **Exploit unavoidable crises or wrong-headed policies as levers to circumvent implementation obstructions and welcome drivers to accelerate change.**

### Tim Benton

I fully agree with you. It looks like Climate Related Financial Disclosures as proposed by the TCFD[138] are going to become man-

---

[138] TCFD: Task Force on Climate-related Financial Disclosures (https://www.fsb-tcfd.org/about/). The Task Force consists of 32 members (https://www.fsb-tcfd.org/members/) from across the G20, representing both preparers and users of financial disclosures. The TCFD is chaired by Michael R. Bloomberg, founder of Bloomberg L.P.

datory in due course, certainly in some jurisdictions. Companies will have to move that forward. And there was a proposal discussed about not climate related financial disclosure, but nature related financial disclosure. Which would then get at issues like biodiversity loss, on the understanding that companies that drive things unsustainably pose themselves as a long-term financial risk for investment purposes. I can imagine both of those doable over a decadal period as opposed to a long term thing, and I think they will incentivise a large scale systemic change. Whether it will be enough, I do not know, but at least it will be something.

# 27  Issues where we could realistically hope to make a difference

### *Tim Benton*

Identify and support change makers, reduce inequality, change the narrative, develop a vision, create a radical transparency for the investment community.

### *Gert van Santen*

Internalize externalities, create a transparent environment in the way things are being priced and promoted. And deconstruction. If we deconstruct something, we create winners and losers. How do we deal with that?

### *Tony Mayer*

We have got to address the whole pricing mechanisms, so that all activity is properly priced to address the issues, you know the carbon footprint, the biodiversity, etc, etc. we need to really move.

### *Sander van der Leeuw*

One thing which I personally feel we could have an impact, if we actually got ourselves organised, because most of you have had very serious and senior positions in the world. And the first aim that I would take, would actually be the opinion industry and in particular the advertising industry. I think we are completely underrating the extent to which the big corporations with all their money, are maintaining our visions and our narratives against all odds. For me, that would be an important one and it is one which I think an concerted effort from our parts to actually engage other people in this, might be something useful, simply because there is not yet any initiative like that.

In the earlier discussion, I put the tech industry, or the information tech industry at least, very much at the centre of what I am aiming at. On the other hand, we are up against a huge power. What is interesting, is that right now, in the US and in Europe, there is developing a very real reaction against that power. In the US, the government is trying to use the anti-monopoly act. There is an increasing amount of lawsuits all over the place. This might be a very crucial time to actually begin to act in this direction, because that would take another ten years before it is to end.

> **Sander van der Leeuw**
> I firmly believe that the total set of feedback loops in the system has created a kind of complexity where there is hardly any measure that could not be useful, but also hardly any measure which could have, on the short term, major effects.

> **Carlo Jaeger**
> The tech industry, is seen by most people as a high technology thing doing wonders. In reality, the business model of the tech companies are advertising, what they are making money with is advertising.

> **Jan Staman**
> When you look at the advertisement of big companies, it is all about green. At this moment, in their advertisements, there is no company who is not green! So the mainstream language — in the Netherlands at least — is that we should be green. And when you are watching the advertisements on television, it is green, green and green.

# 28 Agency and Transformative Action; back away from pessimism[139]

**Hillary Brown**

Martin Rees said (Chapter 3) that we can be technological optimists but there are intractable politics and sociology, social behavior, and gender pessimism. How do we get over the pessimism to the point where we can start to act?

### A.R. Siders: Adding the notion of inevitable change to the narrative

I find this to be a difficult question. For example, I teach climate adaptation and my class is often full of students who think it is going to be an optimistic class because they are hoping that I teach them how everything is going to be ok. How are we going to adapt and be fine? And unfortunately, I always break their hearts. I find that one of the biggest barriers that I encounter when talking to people about managed retreat and their future is that it is very pessimistic. Your town is dying, you will not be able to live here for 30 years, and you must move. It is not a happy conversation. Sometimes it works to try to find a silver lining — but that often sounds grim — that you can move somewhere else and try to find better opportunities, jobs. One thing that seems to start conversations is the idea that change is inevitable, and it does not have to be bad. So, I often ground my conversations in history. For example, in Delaware, I will show pictures of Delaware 100 years ago. "Here is what your town

> **A thought about change**
> Change is one of the few constants in our universe. Yet somehow, that "constant" has become "underemphasized" in the ways Western societies think about (the future of) the(ir) world. These contrast with the ways in which China, through its traditional philosophy, and throughout its history, has used its understanding of the nature of change to reinvent itself.
>
> See also: Hugh Peyman, (2018), China's Change, the greatest show on earth

---

[139] By agency is meant: *The ability (for individuals or organizations) to act or choose what action to take*, https://dictionary.cambridge.org/dictionary/english/agency.

looked like in 1900, here's how it looks in 2000. We assume that in 2100, it will look equally different." Regardless of whether you like climate change, just based on the change we have seen in the past, we expect the future to look very different.

I found that sometimes I can shift to an idea that change is neutral, it may not be positive, but it is something that is happening, and we actually have a lot of agency over that change. Whether or not that is true, I do not know but at least it can become a more hopeful message. It is not really a solution but it is just a thought.

### William Solecki: History

I would like to talk about the historical thread. I look at a lot of cities and I have seen that change is ever-present in our cities, it is part of their history. So, I try to explore this notion that there is an environmental baseline, that cities have been built around that, that they have to deal with climate change, to rebuild and to reimagine, and that this process has in some ways occurred before. If you look at a city that has grown in mid-19th century in the US and you look at London in early 19th century, you see that the ports are going through tremendous disruptions. With that, also new technologies and problems emerge as well as potential solutions. I try to get folks into that mindset of the dynamics of change, to look at the historical context in which significant environmental problems have emerged, like the (air) pollution of London in mid-20th century. How was that solved? What were the issues? What was necessary? Were there issues of water scarcity or quality in mid 19th century New York? How was that solved? That starts to open the mind to think about possibilities and gets some thinking going about this notion of transformative change.

### Nik Gowing: How do we make the unpalatable palatable? (Being realistic)

I think it is wrong to frame the word as positive or negative or I should say pessimism or optimism. I think you should just be realistic. Talking about retreat, I am sitting on the 11th story in London, on a loop of the River Thames. I should not be here because it is inevitable that it is going to flood some time but somehow, I am hoping that it will all go away or will not happen while I am still alive. There is no point being pessimistic about that. I have got to be realistic and make realistic decisions. And that is what a lot of people will have to do, including about imminent climate change, not in 10 years but in 10 months. The words pessimism and optimism are distracting from the kind of calculation that all of us will have to make. It is about how we are going to handle the inevitability, the unpalatability of what we all face. A one degree rise from where Dan talked about yesterday (Chapter 6) is going to release new COVID's that we do not really know about. There is no point being pessimistic about that; you have got to be realistic about that. Similarly with all the retreats Hillary was talking about, we have got to use the right language.

> **Language and actuality**
>
> <u>Localization</u>  Evaluations of programs on sexual reproductive health and population policy on the African continent emphasize that the programs must be local. But who is the local, the international organization that is operating local?
>
> <u>Leadership</u>  In Africa one tends to look at leadership in terms of a political context. Given that political leadership in many places is short-term, leadership is seen in terms of the next election, rather than in terms of the phenomenon that needs to be addressed. Looking at the central role of leadership, we may need to see how we grow this leadership beyond politics.

### Edward Kirumira: Language, history, perspective and sensitivity

We really have not paid attention to the language that we use, especially to the language that is being used as and when we are helping the other. We have to look at the language and most definitely at the context. When we look at the language, I think we need to move from crisis to adaptation and, as has been said, we need to embrace realism. History and language need to be headed to the conversation we are having, that gives us perspective and sensitivity to how we are doing.

### Nik Gowing: we need Mavericks or visionaries

In our work, we have talked about the price of conformity. The conformity which qualifies leaders for their jobs but in some ways disqualifies them from knowing what to do about the enormity of the disruptions that are now underway. First with COVID and secondly with the climate emergency of which Bill Gates is saying[140] is inevitable and that it will lead to many more deaths. We were saying this a long time ago, but it did not get the publicity that Gates got. It is about reframing. We said what you need in decision-making, including in the private offices of Prime Ministers, Ministers, Council Leaders and so on, the maverick voice. Now, if you call someone a maverick, using that word is kind of marginalizing that person in the minds of many conformists. If you then call that person a visionary because they are a maverick as well, that is a positive and that is where we should be looking. People who have got ideas about how it is going to happen, the speed it is going to happen with and then how we work to overcome this inevitable disruption.

> **Origin of the term maverick**[***]
> When a client gave Samuel A. Maverick 400 cattle to settle a $1,200 debt, the 19th-century south Texas lawyer had no use for them, so he left the cattle unbranded and allowed them to roam freely (supposedly under the supervision of one of his employees). Neighboring stockmen recognized their opportunity and seized it, branding and herding the stray cattle as their own. Maverick eventually recognized the folly of the situation and sold what was left of his depleted herd, but not before his name became synonymous with such unbranded livestock. By the end of the 19th century, the term maverick was being used to refer to individuals who prefer to blaze their own trails.
>
> [***] https://www.merriam-webster.com/dictionary/maverick

# 29　Leadership — How do we become more visionary?

*Hillary Brown*

I am just going to quote from Martin Rees again. He says (Chapter 3): *"Spaceship Earth is hurtling through the void. Its passengers are anxious and fractious. Their life support is vulnerable to disruption. But there is too little planning or horizon scanning by the officers on the bridge."* And

---

[140] Gates, Bill (2021), *How to Avoid a Climate Disaster: The solutions we have and the breakthroughs we need.*

I think I want to move on to this topic of horizon planning, about avoiding crisis response and anticipating the emergence, whether it is of disease or of social disruption and how we can be proactive. I recall that Tim said (Chapter 4) that when the system breaks, that is when radical change can happen. But I think we have to have radical change before the system breaks. To achieve that there is a need for planning and thinking that is very bold and visionary in terms.

### Nik Gowing

Hillary, let me just help refine what you are asking, which I think is absolutely right. It is about the word that was mentioned yesterday, leadership. That is what we are looking at and have been looking at. You can have any system in place, but you got to have the human capacity to understand and appreciate the enormity of change and that requires people who are prepared to break with the current model. That is surely what COVID has come to teach us. The way the more visionary leaders were thinking sometime last year is not what they can afford to think now. So, can I help to refine that by saying we have got to think about the human capacity of leaders — and those in leadership positions — to embrace, to identify but then to also act in a different way.

### Alicia Juarrero: Plea for local initiative (bottom up)

I like to make a plea for local initiative because I think that one characteristic of our times today is *mistrust* in leadership. One role of a leader ought to be that of a coordinator.

---

**Alicia Juarrero**

*Somehow, almost all the talks yesterday (Chapters 3–9) sounded very much top-down to me, like: "Here is the policy we're going to implement, here's the way it's going to be done, this is what has to be done. We just need the charismatic people who will be able to do it."*

**Alicia Juarrero: Facilitating local people**

*Cubans are very good at operating with very little technology and very little bandwidth, with very little everything in fact. I was talking to a group and they told me: "We have this incredible project with Johns Hopkins, and they gave us this wonderful software and guess what? You have to be able to be connected or else we will be wasting time and everything". They ended up buying our product (software to assist local communities to monitor and track vector-borne diseases or pathogens).*

It seems to me — and this is not so much for climate change, that is a whole different animal, but for disease anticipation and horizon scanning — that there are tools that will empower people to take the lead (initiative) at a very local level and this can be done in a way that does not require sophisticated or expensive technology. Then the role of the state-wide leader is to coordinate, facilitate and identify those initiatives so that we can amplify — not top-down but bottom-up — the fluctuations we are looking for.

You have to know the needs of the locals and then support them with technology that fits their capabilities and local circumstances. For instance, a cellphone, that let you import, collect, and export data (to the cloud) without having to be connected. Then the data is there and, when you get back to the office, it all comes in through the cloud so you do not have to worry about hurricanes wiping it out. The locals can visualize their own town through maps and bar graphs, and they can see: "Look, the epidemic started here, and it'll move there. Let's make sure we vaccinate people living on the leading edge. We need to integrate the weather, you have to integrate social determinant data and everything". You cannot ask people to understand the technical terms, but we have to be able to figure out what minimum scaffolding[141] folks need so that they can click a button and see — here is my situation at the zip code level in the United States, or at the municipality level anywhere in the world. That must be scalable so that the CDC[142] and the health system can see the whole picture, while the data collection and monitoring is in the hands of the individual locality. I think if we can empower people that way, we have got a chance. Forget top-down, this is what needs to be done.

### A.R. Siders: Tension between maverick leaders and bottom-up

This is a really interesting point. I think there is a real tension between needing a maverick leader and needing bottom-up. Many localities

---

[141] A scaffold is like an infrastructure that you can provide top-down but that is not an edict or dictative, but one that folks can climb up on their own.

[142] CDC: Centers for Disease Control and Prevention, https://www.cdc.gov/.

do not have a maverick leader, so if we wait for such a leader to come forward, we will have no change. And so, the pace of change or innovation is slow, we see this in the US. Which towns embrace managed retreat and which towns do not? The towns that embrace managed retreat tend to be larger, denser, wealthier and have very vocal leaders. And there are a lot of towns who want to retreat but they do not have the capacity or leadership to do so. Even though we theoretically have a scaffold in place, it still does not happen.

There is also an interesting tension between wanting that bottom-up and recognizing that many of these bottom-up solutions have to be coordinated. Because one town's decision to retreat or not retreat has implications on all the towns around it. A town can make its own choices but those choices will affect everyone else so there is an argument to be made that if not top-down then at least top-level coordination. So, I think this is a really great point that we need bottom-up

**Bottom-up and top-down**

*Especially coastal areas, which is primarily where we see retreat right now, struggle with "making your own choice (bottom-up) and top-down coordination".*

*"If I don't retreat and build a giant wall, our town will be destroyed instead. Ok, I made that choice because this was my grassroots bottom-up decision".*

ownership to empower people and yet there is this real need for coordination and scaffolding that does not seem to be present at all.

### Hillary Brown: Top-down does not have to be so heavy

This tension maybe partly resolved by emphasis on localization and developing inter-champions, while being wired and connected to best practices around the world.

### A.R.Siders: Transfer of best practices is hard

I agree yet I am concerned about the ability to transfer best practices. For example: the United States has been running a managed retreat program for 30 years, a formal and federally funded program, but to this day, I will talk to people in FEMA[143] and they will say it does not exist. I will then point them to their best practices manual from a

---

[143] FEMA: Federal Emergency Management Agency, https://www.fema.gov/.

decade ago and they have no idea that these things exist. So, a level of education and outreach is required to transfer these best practices. I think it is there, but I have become skeptical about it, because I see so much effort in transferring best practices yet it is not happening. And I am not sure how to actually raise awareness about it. Maybe we need local leadership to that end? Or some top-down leadership to spread the word?

# 30 History and agency

### *Edward Kirumira*

Yesterday, there was a talk about politicization of issues. Sometimes, you do not want to politicize issues. Take for example immigration on the African continent. Historically that has always been there. Actually, there is more intra-continental migration than inter-continental migration but the crises tend to make us look at crossing the Mediterranean rather than the intra-African migration. So the challenge we have to address, lies where the problem becomes us rather than the other.

When we get people into a mode of being helped, they actually become dependent and sometimes take advantage of us. They promote the narrative of dependency for their own benefit. I have seen programs that are helping "helpless individuals" and when you look at these helpless individuals, they have the best mansions. So, I think we have to be careful not to create a helpless situation that becomes counterproductive. In dealing with the African or Asian or Southern, we need to get to the stage where we demand that they are responsible for what they do rather than saying that it is something else and therefore they need to be helped. There has to be agency.

### *William Solecki: How is the present different from the past*

How have we solved problems of the past? What is the scale of the issues and how and why is this going to be different? What are the tools that we need specifically. A lot of it is embedded in the issues of governance, trust and equity, which we haven't really discussed yet.

> **SR15 is an IPCC special report** *on the impacts of global warming of 1.5°C above pre-industrial levels and related global greenhouse gas emission pathways.*
>
> *The report makes quite clear that we have the technology to address the issue of climate change but that the big challenges that come up consistently are financing, governance and all sorts of cultural issues.*
>
> For more information, see https://www.ipcc.ch/sr15/.

And then there is the financing question. Where is the money coming from? We can have a lot of mavericks but if they do not have the resources available to them, they cannot do anything. So the question I keep coming up with is how and why this situation, at different scales and questions of governance, is different from the past?

### Dan Brooks: Moving away from danger

Looking at how human beings coped with climate change in the past — until about 11,000 years ago — it is noticeable that whenever there was a climate change event that adversely affected the human population, they just left, away from danger, discomfort and changing conditions. That is one of the reasons migration has been so deeply embedded in Africa. But, in the last 11,000 years we, (humanity) have anchored ourselves in place and we have changed the terms of reference ourselves. That is one of the reasons why we have many of these problems of climate change. The issues of the city and disease, of cities and sea level rise, of cities and warming are all tightly interconnected and a lot of it has to do with the fact that human beings have changed their own evolutionary trajectory in many ways.

# 31   Financing prevention and long-term planning

### Hillary Brown

How do we convince people that pro-action is cost-effective? And what are the ways to engender financial cooperation across diverse parties not typically working together?

## Edward Kirumira: *We need to change our mindset*

I think raising funds is not really a big problem. The real issue is preparedness and prevention, rather than curation or solving. For example, we have not taken advantage of the lessons learnt through our various responses to HIV and AIDS. Unfortunately, it seems that we are not learning from what COVID teaches us. We are just looking for money to develop a vaccine to block this thing, rather than say that this teaches how to prevent other pandemics. We are thinking in terms of solving, not in terms of preparing and adapting. So, how do we start financing for prevention?

We just have to change our mindset to say that. I think that if we do, we will find that the finances are there. And that we would be able to make some progress.

## A.R. Siders: *we have no language for a targeted shift in investments*

In discussing how to make prevention and long-term planning more cost-effective, we do not have good language for talking about disinvestment or areas where we are not going to invest. This is more relevant to the managed retreat conversation. In that conversation, there are always investments for people to move and go but there are also places that we are not going to invest. "We're not going to rebuild that bridge, we're not going to build that road because we think that everyone should be moving away. We're not going to provide the local government money because you know, it's going to be a bad investment". And we have to figure out how to do it equitably and we have not talked about the equity issue that is so difficult. We do not even have the language to talk about non-growth or de-growth in areas. I do not know, I am struggling to even talk about it because all the words in this field are so charged and we do not have a good language to talk about that kind of targeted shifting in investment.

### Hillary Brown: Our consumption pattens define us

In developing the language, how do we differentiate the industrialized world and the developing world? When we talk about finance, the wealthy countries should really be contributing to development and redevelopment, to enable leap frogging in developing the nations in the developing world. I recently read an article about when you throw money into some of these countries, corruption is rife and it will never trickle down onto its intent. So, I think that finance, just throwing money, is a really loaded issue.

### Edward Kirumira

One of the promising developments in some parts of the African continent are the start-ups. Nairobi has probably the largest number especially those that are based in mobile and telephone. Most of this development is driven by young people. I think this helps enormously in bridging rural and urban divides, and all these different kinds of things. There is an opportunity there, that is really going back to the localization of

> **Alicia Juarrero: local bubbles**
> *Find and amplify the local bubbles, that is a lot cheaper than spending 3 days in Geneva, trying to figure out a program that costs an incredibly amount of money and that does not work as effectively as when you amplify the local capacity. That capacity is already there and operating.*

home growth. If you look at some of these 4G phones and the impact they are making, for example with regards to all kinds of warning systems, there is an opportunity that one can use to channel the financing. That is of course if the big companies do not buy them out. Those are the sort of foci that one can easily build on.

# 32  Vested Interest — absent political will/nerve

**Nik Gowing**
*Political nerve is a better word than political will. If you look at the climate emergency and the need to decarbonize, it is going to take nerve for example in the UK to tell 26 million households that they got to change their gas boilers. It is not about will, they got to do it if we are going to get to net zero as promised.*

**Hillary Brown**
*How can we effectively* **communicate a** *sense of urgency in those with the power to say "yes?" With regard to political will, we need a social tipping point. If we get the populace educated enough to understand what is happening and changing their hearts and minds, we should see the politicians following.*

### Nik Gowing: A private and subjective impression

I would suggest that in the UK there is a significant change in recalibration of political nerve going on when it comes to the climate emergency and sustainability. I do think that in the UK, you have seen a significant shift in the last few weeks[144] out of determination on the climate emergency. The tagline for COP26[145] is essentially: *make net zero a commitment, then you are in. If you are not in by November, you are out and your shareholders and stakeholders will have to make a decision.* That is a significant focusing point and it will take political nerve to face and implement the consequences of that decision. I think what we are seeing is the potential emergence of an international agreement to write off debt because no one is going to buy electric cars without some kind of subsidy. No one is going to sub-

---

[144] January 2021.

[145] COP26: 26th UN Climate Change Conference of the Parties, will take place in Glasgow in November 2021, https://ukcop26.org/.

scribe to the kind of net-zero economy without significant financial encouragement, and it has got to come from somewhere (at a time that the UK is already about 300 billion pounds in debt because of COVID). There is actually a precise way of putting it, which is *political nerve* and *determination*. The will and the nerve will come when the right decisions are taken because that is the only way forward, even if there is a humongous amount of debt which eventually will be written off. We are not there yet, but we are moving in the right direction.

### Hillary Brown, a small example of political nerve

Another example of political nerve is given by the Mayor of New York City (de Blasio), or actually the City Council that embraced an aggressive building retrofit law, holding landlords' feet to the fire. I think it took nerve to shake up the real estate community in a local town (even though we may not see the progress we want). To the point of financing, what that act is lacking is an efficient financial incentive to make the changes happen. Eradicating debt may be one way of looking at that.

**Martin Rees** quoted Jean-Claude Junker saying: "Politicians know what's the right thing to do, but they don't know how to get re-elected when they've done it." I think that is changing quite rapidly, not in a way which is irreversible, but it is beginning to move into that direction.

### Nik Gowing: Change is coming

You may not agree with Larry Fink[146] or you may not like him but what he wrote in his CEO letter[147] is important. He happens to be the biggest hedge fund investor in the world. But when he says that many people are coming to him now and saying: "Don't invest in anything that is not climate net zero", that indicates an important change. It is affecting the pension funds, it is affecting everything that has

---

[146] Larry Fink is an American billionaire businessman. He is the chairman and CEO of BlackRock, an American multinational investment management corporation.

[147] For more information about the letter to CEOs, see https://www.blackrock.com/corporate/investor-relations/larry-fink-ceo-letter.

a legacy connection to carbon-generating stuff like coal and so on. You are seeing this from BP and from Shell, you have not seen it from Exxon yet. But what you are seeing is language and action. If they mean what they say, then these are significant changes. And if anyone sees large investing organizations like BlackRock saying: "Don't invest if it's dirty stuff", you have to believe they're doing that and that Larry Fink is not lying.

> **Shifts in the last few weeks**
> – Biden taking over from Trump.
> – China's commitment to net zero.
> – A significant shift among the corporate sector, in language and commitment. "We're less interested in our shareholders and more in our stakeholders. Who's going to buy our output? Who's going to buy our product? The next generation doesn't want to buy things that are dirty and are not grown sustainably". It is not majority yet, but the World Economic Forum mentions 12,000 major companies in the corporate sector, even if some are greenwashing.

What you are seeing is the potential. According to the World Economic Forum, there are 395 million jobs being created from green. Yet, in Eastern Europe there is still suspicion about the great Green Deal. "Why do you want to get rid of old-fashioned industries?". That is where it comes down to political nerve — selling the ability to go green to generate an even better economy. It is going to take the political nerve, not of Orban in Hungary but of Biden and John Kerry.[148] And already you are seeing this significant pushback against the Republican party in the US, who are anti-green. That is a significant change.

### Daniel Brooks: The pathogen perspective

Pathogens do not recognize national borders, races, religions and economic systems.

So, one of the things that I have been saying is that if we want to be more proactive and preventive about emerging diseases we are going to have to cooperate with our neighbors, and especially with neighbors that we do not like. That too requires political nerve. When I was in a Research Institute in Hungary I had the following discussion with a former member of the Hungarian government:

---

[148] US Presidential Envoy for Climate.

I:    "What is your government going to do to help Romania with its potential sea rise problems?"

He: "Oh well, nothing. Why should we help them?"

I:    "Well, when the Black Sea rises 7m, where do you think the Romanians are going to go?"

He: "Well, they'll just stay in their side of the mountains."

I:    "When in the last five thousand years of history has that been true?"

Whenever you talk to neighbors or somebody connected politically, there is always a point of disconnect. To (re)connect and search cooperation also requires political nerve.

> Select people with proven high-level leadership qualities and politicians with nerve to direct adjustment of existing macro-policies and regulations and enable regular, transparent monitoring of their implementation and effects to instill integrity, trust and respect in this complex political process.

# 33    Linking bottom-up and top-down to catalyze action

### William Solecki: Can we find a sweet spot?

"Are there effective ways to merge bottom-up and top-down processes? A lot of us know about Charles Vorosmarty's[149] use of the phrase "sweet spot". Where can the most effective linkages be made to catalyze action?

---

[149] Charles J. Vorosmarty is the director of the Environmental Sciences Initiative and the Environmental Crossroads Group at the Advanced Science Research Center at the City University of New York.

### Alicia Juarrero: Coordination is the link

I think, the link is coordination. Rather than dictating or claiming to be the solver of all solvers, top-down simply coordinates and amplifies the bottom-up. In turn, that will change the top. It is a feedback loop of enabling constraints — you enable from the bottom(-up) and you set the perimeters from the top(-down). That is the link — it is not an either or, it is a dynamic loop.

### Daniel Brooks: An example

In Costa Rica, the conservation areas in the rural regions are sources of biodiversity and of income from tourists. The relationship between sustainable use of biodiversity and socioeconomic development in those areas are based on that idea. It works through parataxonomy.[150] You go to the villages surrounding the conservation area. In those villages are people who do not have the formal education but who know the local diversity very well. You identify those people and then you build up that capacity. You do not send them to a university, but you say: "This is what you know, we need this information and we're going to amplify that capacity". Those people then become the essential technical support for international scientists who come to do research. At the same time, those people have families and friends in the village. Everybody sees that their neighbors are making money by helping and having jobs in the conservation area. They see "Oh, there are more tourists coming. My restaurant is making more money." They begin to see real socioeconomic benefits from those activities. For example, the team in the conservation responsible for controlling wildfires is made up of people originally arrested for setting fires for poaching and they are now making more money putting the fires out. It can work but it is very time-consuming and you must have people who are willing to invest in identifying that capacity and then

---

[150] Parataxonomy is a system of labor division for use in biodiversity research, in which the rough sorting tasks of specimen collection, field identification, documentation and preservation are conducted by primarily local, less specialized individuals, thereby alleviating the workload for the "alpha" or "master" taxonomist, https://en.wikipedia.org/wiki/Parataxonomy.

amplifying it and then benefitting from it. Costa Rica is a little tiny country that nobody wants because it has very little mineral resources but it is a special place where you can see that this kind of thing could work.

> **Focus climate related strategies on the critical duality of the role of high level (national and global) policies, while providing maximum flexibility for the sharing of power and implementation of multiple, parallel actions at lower levels aimed at converting unsustainable production, trade and human settlement practices.**

### Hillary Brown

Right, so what we are saying is that there are champions for change that can emerge, that they are just naturally latent in communities who somehow get up enough nerve to assert themselves and it is those individuals who can tap into the higher level or the higher level can tap into those individuals to tap into the proactivity. So, that answers William's question on how to merge bottom-up and top-down processes.

### Alicia Juarrero: Another example

In the Dutch Antilles,[151] we worked with the surveillance team of Vector Analytica with local people. We worked manually on paper by providing the tools to enable them to quickly and easily input the data and weather on the cell phones they already have. And then a question came from the fishery people: "Can we use your app to discover information about prohibited fishing techniques that are being used, and is there a way we can report on that?". And all of a sudden, you have fisheries getting integrated into vector technology and simultaneously, "you know what, we seem to have rodent problems in the islands and can we use that app to integrate". So the question is: is there a relationship between the mosquitoes and the rodents. And so the whole thing scales up and integrates. Instead of having a

---

[151] Aruba, Curaçao, Sint Maarten, Bonaire, Saba en Sint Eustatius, are islands in the Carribean.

project for studying rodents, with someone writing a paper that will get published and nobody reads, this is an ongoing standing platform that you can build on. All you have to do is to be responsive: "Yeah we can do that". And it is not overly expensive, it is costing a nickel, 5 cents per person. You can create something that the whole country can use, from the top-down and bottom-up, and it can integrate regionally because Bonaire and St Eustatius are up near Puerto Rico.

# 34  The case of managed retreat[152]

*During the webinar Managed Retreat (Chapter 7) and Infectious Diseases (Chapter 6) were discussed in one joint session. Both topics deal with consequences of climate change. For both extensive action plans have been developed that now face the hurdles imposed by stumbling blocks. Partly because of the backgrounds of the participants, this discussion focused mainly on Managed Retreat and it so happened that it entailed and made explicit many of the aspects highlighted in the other discussions during the webinar. Managed Retreat is therefore reported here in slightly more concise and structured way than the previous discussions.*

### Hillary Brown

What are the barriers and opportunities for managed retreat? One proposal is that we can solve multiple problems if we look at hinterlands. How do we address equity issues that arise in this context?

### William Solecki

Interesting social science work alludes to anthropological and archaeological work — where and how do people make decisions about

---

[152] Managed retreat involves the purposeful, coordinated movement of people and assets out of harm's way (https://www.earthsystemgovernance.org/publication/managed-retreat-in-the-united-states/). See also:
- Guest Essay by Katherine Mach and A.R. Siders in the NY Times (16 July 2021): *Is Your Town Threatened by Floods or Fires? Consider a 'Managed Retreat'*, https://www.nytimes.com/2021/07/16/opinion/managed-retreat-climate-change.html.
- *The case for strategic and managed climate retreat*, https://science.sciencemag.org/content/365/6455/761.

migration. Geographical and hazard literature suggests where people have gone. In the urban context, they move to the next community or the next spot. Long-term projections on where people will relocate, indicate that Florida people may move to other parts of Florida, but increasingly will move inland. Equity (being fair and impartial) is the number one issue regarding vulnerability and processes of managed retreat. New York is trying to promote resiliency, but local communities see this as a gentrification[153] of their community and a way of getting them to get out of there.

## Framing and implications of the term "retreat"

### Hillary Brown

There are two kinds of retreat: voluntary, where people elect to move to the next location, and forced, a wholesale abandonment of the community, moving everybody elsewhere.

### William Solecki

People in all these high-risk communities and sites forget that this is not a retreat but an advance (improvement). The US government made conscious decisions to rapidly develop high-risk water sites and they sort of created a great big apparatus to enable that. We have a long history of advancements in different contexts but not of retreat. Wording is important because people respond dismally to the question of retreat.

### A.R. Siders

Major literature does not cite enough on migration, on what influences people's decisions and what their experiences are. While not

---

[153] Gentrification: A process in which a poor area (like part of a city) experiences an influx of middle-class or wealthy people who renovate and rebuild homes and businesses and which often results in an increase in property values and the displacement of earlier, usually poorer residents. For more information about the Urban Displacement Project, see https://www.urbandisplacement.org/gentrification-explained.

all migration is seen as a negative, managed retreat is almost always portrayed as a harm. How can that be? Fundamentally in the US, people do not know if man-aged retreat is a good or a bad thing. If it is good, we should prioritize providing funds to communities that have been marginalized and are experi-encing injustice, because we should be helping these com-munities to get away from the risk. But if it is bad, we should not be doing managed retreat in those communities because we would be inflicting greater harm on them.

> **Is managed retreat good or a bad?**
> We did a study that showed that there is a disproportionate effect on lower-income neighborhoods in the US, that also tend to have a greater percentage of racial minorities. Is that good because they are getting the help they need or is it bad if these communities are being disrupted and losing their history?
> Some officials say we need to prioritize low-income homes **because the people there need the help the most** and we need to help them get out and we absolutely do **not** prioritize low-income homes **because of the homeowners**. All you do is offer to buy the homeowner's home. The homeowners still make the decision. There is a <u>huge debate</u> on how top-down or bottom-up this is.

Deciding what it means to be equitable is in itself a very difficult problem. You can point to clear inequitable examples in the Philippines and Lagos, where people are being relocated at gun-point. Or take clearing space for investments in high-rise development. If involuntary, we can point to those as being inequitable.

But other examples are messier. For instance, when people feel coerced because they feel it is the only option open to them. Like when their options are to rebuild in a town that can be flooded repeatedly or to sell their home. They do not want either of these options, so they just pick the least bad option. It is really difficult to make people see reality and convince them that this is actually the only option. And if they must pick from a limited set of options they do not want, how voluntary is that? Is this an equity problem? At what point is it "a something else" problem? These are incredibly complicated issues that we do not have good answers to. And I am using "We" in a very generic sense because it is not even clear who the "we" is that should be making these decisions. We can say that everyone can make their own choice but it becomes very difficult very quickly.

**Hillary Brown**

What if we create a lure and reframe managed retreat as a positive? Is it a realistic possibility to match vulnerable communities with communities that need to be revitalized. Is it in any way viable to portray this as an opportunity versus a loss?

**A.R. Siders**

I think it is possible to frame this as an opportunity. One of my favorite examples is in the early 1980s when an entire town in Wisconsin (Soldiers Grove)[154] was relocated away from the river. When they made the relocation, they not only revitalized their downtown, but they also put in their own nursing home because of their ageing population, and they wanted to attract elderly people from the surrounding countryside to their town. And they went completely

> ***From the NYT archives, June 29, 1987***
>
> *Soldiers Grove, population 622, mandated solar heating when downtown businesses were moved to escape the flooding on the Kickapoo River. For the most part, nobody is sorry. The town fathers had decreed that old businesses move away from the river and that new business structures be solar heated. Ed Herbst, a 69-year-old tavern owner, [...]: "We were flooded out seven or eight times in the old town. This town had to make a decision. Would we move and rebuild, or just let the town die? We weren't about to let it die." [...] The village began with Joseph Brightman's sawmill on the banks of the Kickapoo in 1857. A hydroelectric plant was built at the turn of the century. Meanwhile, the river valley was losing its ability to absorb rain and melting snow as land was cleared and timber cut. Floods came in 1907 and by the mid-1930's water was damaging buildings almost every spring. [...]*
>
> *In 1977, the village paid $82,500 for the relocation site, but its business district was still in the old downtown area a year later when the Kickapoo smashed through earthen levees and swept down Main Street, leaving damaged, mud-filled businesses. After that, the town decided to relocate all businesses. [...] Twelve homeowners in the flood plain were given permission to flood-proof their houses rather than relocate. Most of the village's homes, however, already were located above the flood area.*
>
> *With energy costs expected to soar, the village board decided than each new business building must receive at least half of its heating energy from the sun.*

---

[154] For more information, see William S. Becker (1981), *The Making of a Solar Village a Case Study of a Solar Downtown Development Project at Soldiers Grove Wisconsin.* https://www.amazon.com/Village-Downtown-Development-Soldiers-Wisconsin/dp/B004OUKY8A.

solar, which in Wisconsin in the early 80s was radical. So, they became known as Solar Town and it led to a really good economic resurgence of a town that was dying beforehand.

On the flip side, some research suggests that the town is divided. The people who moved, loved it, and thought it was fantastic. The people who did not move were unhappy that half their town moved. They did not want to move, and they did not want anyone else to move and so when everyone else moved, they were very unhappy about it.

And that raises the question: What do you do if you cannot make everyone happy? But I do think it is a clear example that shows that there is an opportunity and the people who moved saw it as a benefit. The economics clearly show that it is a benefit, the population growth suggests that it is a benefit as well. But we need to bring in other things because it cannot just be about relocation. It also must be about where the housing is developed, where the jobs are going and where the new school is built? There must be other components and frankly our government systems are not well-designed to do that kind of holistic and long-term planning.

## Barrier Housing equity

### William Solecki

Housing equity is rapidly emerging as a central issue, a looming and rapidly accelerating crisis. You see loss of housing and housing opportunities in these high-risk areas but no proper building for lower-income houses in a lot of these urban settings.

### Nik Gowing

What is happening in the UK after three years of flooding, is that insurers are not insuring houses anymore. So already you are beginning to get retreat. You are also beginning to get people marooned in houses because they cannot afford anything else. So, if you are talking about the enormity of the

> **Using the right words**
>
> *We should not use the word "retreat"; it suggests going backwards against adversity. We should use "relocation" or "re-siting".*
>
> *And we should not use the word climate "change", but climate "emergency".*

adversity of climate emergency, we already see this manifested in people's bank balances or in the fact that they cannot insure at all. In the UK it has become a kind of life motif: can I get this insured or not? Does this have value or not? What do the banks say about mortgage, et cetera? You already begin to see it. But it is becoming an enforced phenomenon as opposed to something that people could have envisioned themselves.

### A.R Siders

In the US, insurance companies pulled out. They say they will not insure these coastal properties until the federal government insures them.

### William Solecki

We have not developed a reinsurance act for the States. And banks are pulling back from high-risk sites, so it is harder and harder to get mortgages for homes.

### A.R. Siders

On the larger scale (in the US) we also see that companies are starting to incorporate climate change into the credit rating for a city. This may have consequences for a city that wants to finance a new bridge or finance a new school. We have already seen a number of such plans being rejected because of climate change. On the one hand, this may be an amazing driver for creating political nerve (*"If it hits you in the wallet, maybe you will act"*). But the concern is that if cities and towns are already struggling to finance climate adaptation projects and they cannot get finance to do anything (like build new affordable housing at a safe place), then how do they do both at the same time? And we do not have language to say that a town is facing de-growth or should die, and that it is not going to be saved. All the retreat that we have been talking about is

> **We do not have the language**
> *We got cities like Detroit and we do not have a way to talk about the fact that maybe this town is so at risk, that it should cease to be a town. Our process should not be to invest in this town and keep it there, our process should focus on how to invest in this town so that it can transition away from here?*

picking up a town and moving it so it still exists somewhere or about moving homes from the edges of a city that still exists. We have not gotten to the point where we say that this town will become part of a different town or it will just not exist anymore. I think we are getting close to that, we begin to see examples, but we are not ready to talk about it yet.

### Nik Gowing

We are beginning to see this in London. There are parts of London that do not have to exist anymore because there is no one in it. So already we are beginning to see a push back in the insurance market because why insure something that has no value anymore.

### Coleen Vogel

From my country's (South Africa) view the complex informality and informal settlements make this all seem really nice in theory, but in reality?

### A.R. Siders

Informal settlements make these issues particularly challenging, and I think that is a major reason why the idea of retreat has to go hand in hand with new housing development.

### Coleen Vogel

I guess what I am struggling with is that you are all talking about towns that are well organized and where everything is very neatly in place in planning terms. Where I come from (Johannesburg, South Africa), town planning and management are messier. We may not even know how these things are emerging and so then how would they apply?

I stay near the Jukskei River that floods regularly. We have tried to be proactive through disaster risk reduction and actually suggested people to move. But as soon as people are told to move, a whole set of issues arise. Land is everything and the people are probably migrants coming from outside the country and if people do move, the land (or settlement) is taken by 5 other families who are trying to get work. So I guess I am really struggling with this notion of having everything so nicely packed in a box but how does it play out in the Costa Ricas, the Johannesburgs or the Brazils of the world?

### A.R. Siders

The Philippines probably has the best example that I can point to. They have done quite a number of relocations of informal settlements after major hurricanes. To be fair, with mixed success, because people just return over and over again. Context matters and raises a whole host of different challenges.

## Opportunity: Retreat as mutual interest and cross-border

### Edward Kirumira

I would like to make two points. First, an example where some form of retreat is happening in a different way. There are some studies coming out of Luanda, the capital city of Angola, where they are trying the concept of urban development around interests and professions. Instead of developing urban areas in the traditional ways, this particular area is built for artists and artistry and museums and things like that. In other words, they are trying to move beyond ethnicity and beyond traditional divisions and work with individuals around common interests. And they are building walkways, even in high-rise non-accessible areas, opening up old neighborhoods that people can walk through even if they do not live there. It is a model one can look at. Retreat is being seen in terms of interest and not as spatial retreat. It is not hugely successful, but I think some areas in Luanda have become really interesting.

The second point is that perhaps we have been so much forced into this mode of making everybody happy we have ended up not making anyone happy. So how do we proceed? Throughout history, there has never been a time where everybody was happy. And we should not equate equity with equality, because then we get stuck with politics and sensitivi-

**Pathogens**

*Pathogens offer a great opportunity to work cross boundary because pathogens do not know boundaries. To learn how to prevent outbreaks of HIV, or the Marburg and Ebola viruses requires cross boundary studies and cooperation.*

ties. I think we need to look for challenges that cross borders. For instance, in commissions, like the Lake Victoria Basin Commission, and water bodies, we were working on finding cross-border solutions. There are opportunities in regional bodies. I think we have to let go of some of these notions of Africa, Black, Coloured, White, that kind of thing. Following such approaches (moving beyond ethnicity into areas of common interest like they do in Luanda or taking up cross boundary challenges like in the commissions and waterbodies) also makes it possible to avoid building on problems that were shaped by history.

### Alicia Juarrero

And it solves an additional problem, namely this notion of people feeling that they are losing their identity once they move. If it is regional in that sense Edward meant, the loss would not be that drastic because people still feel part of this particular group.

I hoped for a while that after the threat of Brexit some regionalism would emerge in Europe, where cross-border regions would have been developed, that would be more effective, more sustainable. Then people would not be as upset about moving because it would still be within their family ancestry of region. And that is a huge issue when it comes to moving people physically. As it is, I do not think managed retreat would have ever been thought of by a European. I think Americans move a lot easier.

### Hillary Brown

I think that this additional regional dimension is a very smart and important one. It also aligns with potentially ecological phenomena, which makes adaptation and dependence on local communities more viable. Working in a watershed for example. We can do a lot of good in restoring a watershed to its former functionality for irrigation. It is kind of border busting in a sense that it gets people to think outside the normal hierarchies of city, county, state and federal. So, drawing new boundaries and looking at regions of mutual interest could be very promising.

*Willam Solecki*

Within the US context, the federal government has made conscious decisions about how to create metropolitan scale suburbs so that there are all sorts of these incentives, everything from highways and new types of development. But then the problem was that the poor were left in cities. Every policy has implications — who gets left out?

We heard the example of Angola, these districts where people are coming together in a sort of new context. The question of refugee rights is complicated in terms of the places they are coming from, and the receiving areas and there is some tension internationally regarding this whole context. So, I wonder, is there a scale that indicates where and how relocation processes work better and are not front loaded with problems?

## Moving economic opportunities to rural areas

*Dan Brooks*

In the US, moving economic opportunities to rural areas turns out to be a bit messy because a lot of rural parts have been affected by decline. But we have reached a point where the time to do something before events overwhelm us, is running out rapidly. And we have to be realists. So, we need to act before we are forced to, by providing economic opportunities for people to move from cities, that may be at risk in 10 to 50 years, to rural areas. The difference between property values in the city (high) and in the rural areas (low) will provide the money for moving. To give people reasons to go, we also need to move corporate interests into these areas, so there will be economic opportunities for those who move.

> **Decline of rural areas US**
> *The Great Plains is an area that is expected to experience a 30-year drought cycle, extending into 2050. The current projections are that if the drought is at the low-end of the projections, the US will lose only 30% of its food production but if it is at the high-end, the US could lose as much as 65%. Increasing the number of people in areas that are under greater stress for reduced food production will create another kind of conflict.*

### Hillary Brown

Whatever entity is in the position to judge long-term climate viability, needs to identify those regions and then we have to figure out the levers to lure people to move there.

### Dan Brooks

I agree with Nik that we must understand that, as unpalatable as it is, we are facing some real existential threats, primarily to technological humanity. I mean there are people living up in the Amazon who are going to be just fine if "civilization" collapses. But, we are very close to the point that if we do not take preventive action on something proactive, events will overwhelm us generating the most inequitable of all crises.

**On luring people to move**

*It is in our biology to try to leave a place where we cannot survive, where we cannot feed our children. But, will it be possible to say that your children are going to be in great risk in about 10 years but here is an area that offers opportunities for you to move to and protect and feed your children? Some people, automatically, will choose to wait. And that is also part of our history. Some people choose to wait until it is too late.*

### Coleen Vogel

When scientists study extreme weather, like massive heat waves, they see it in discreet little boxes. It requires science fiction writers to show what it means for us in the real world. So, where and how do we start to shift proactively? I have been working with the city of Johannesburg, trying to do this and they keep very politely saying that it is not on their scorecard right now. I hear the point about the next generation and intergenerational justice issues but it is hard, even to convince my own metropolitan municipality officials to take this on board.

## Embrace unconventional thinking

### Edward Kirumira

Last year, we (the program committee) agreed to fund a young African scholar to study how stand-up comedy works in informing people and transferring knowledge. It touches on what you Coleen said about science fiction writers, and it addresses this whole issue of stepping outside of ourselves as scientists to see things that can actually help us to see the future, rather than just dealing with the present. I think science fiction writing and stand-up comedy help. Stand-up comedy plays a big role in political transformation in the continent because it provides that space where you can say what people usually cannot say as a scientist or a political scientist. I think perhaps we also need to be unconventional in the way we are thinking about solving these issues.

### *Hillary Brown*

That is a really strong point that brings us back to Nik's point about the idea of having nerve. It is to transcend the conventional thinking and to launch into a space that is safe to talk about these issues. And the arts is clearly one way of doing it.

## *Connection between relocation and maladaptation*[155]

*The term maladaptation originates from evolutionary biology. There its identifying characteristic is the absence of evolvability. It is a dead end. Within the context of climate change negotiations the term was redefined by the UN IPCC as: "…an adaptation that does not succeed in reducing vulnerability but increases it instead".*

### *William Solecki*

When I talk to my engineering friends about managed retreat, they say: "You have to do the calculus of the life cycle assessment, like how much is that going to cost in terms of carbon emissions and all the related points that come with that." In other words: calculate whether the change brings more or less burden than the situation of no change (more burden means the change is maladapted).

### *Hillary Brown*

Related to that is the question what we do with all the stranded assets? As we retreat from one area to the next, what becomes of them?

---

[155] For more information, see https://wedocs.unep.org/bitstream/handle/20.500.11822/27545/Frontiers1819_ch5.pdf.

### A.R. Siders

It is really important that all around the world, we are thinking about: if not retreat now, at least take steps now to enable future retreats. Like, if we add more infrastructure in a place that would have to be abandoned, this is only complicating these issues. And that requires an admission that the place may have to be abandoned in the future. Just imagine the difficulty of convincing Phoenix that they may not exist in 100 years and to tell them not to build roads or repair minimally.

This is a sticking point but it is important because we are raising all these really great points about how we are going to handle maladaptation and the new building infrastructure and how do we reuse assets. But it also comes with this, how can we convince people that we have to make these decisions today in order to make future retreat easier.

### Hillary Brown

Well, that is the kind of forward-thinking we need to do. We do not have to tell people that we are retreating today but we can make plans accommodating them for the future.

# Final reflection

Humanity, especially technological humanity, is facing an existential crisis that may well come to a head in the next two generations. We can no longer hope to stop global climate change, perhaps not even to slow it down much. At best, we will face massive disruptions in almost all aspects of life. At worst, we could lose most of our technological infrastructure. Realizing and accepting this threat means we have a clear and certain mandate to move from aspiration to operation. We must immediately begin to implement effective action plans to cope with the changes that are coming at us at an accelerating pace.

The five categories of stumbling blocks identified in this discussion are large, immobile and powerful agents of social management. They are largely instruments of a world whose global climate is stable, so they are ill-equipped to cope with the realities of accelerating global climate change. The first movement toward effective action, therefore, was to decide how to cope with those stumbling blocks.

The participants in this exercise have largely come to the conclusion that we must avoid conflict with the stumbling blocks, rather than attempting to topple or alter them fundamentally. This is because at best such conflict would be a waste of time and resources, and at worst, would not make any difference, leaving humanity in an even more precarious state than presently.

The participants also seem to be largely in agreement that the most effective way to avoid the stumbling blocks is through grassroots action. Grassroots initiatives are sensitive to local diversity of needs and desires. This allows us to avoid the pitfalls of *one size fits all* planning, replacing one inflexible operation with another at a time when flexibility is paramount. Grassroots initiatives allow us to take action quickly, because the actions are done at a local scale where essential services like circularizing economies can be done in the most cost-effective manner. Even better, there is a global collection of grassroots initiatives — some more effective than others — that we can use as templates. We do not have to reinvent the wheel.

The focus on grassroots initiative does not mean there is no role for the systems that are currently stumbling blocks to effective action. Far from it. The relative immobility of the institutions representing the stumbling blocks can become assets, providing stable and predictable support for grassroots aspirations, reinforcing the outcomes of action plans that work. In this way, we may hope to restore trust in those systems at a time when such trust is waning.

To paraphrase an old saying, "Think globally, act locally...Begin NOW."

# Annex 1: Speakers, invited participants and organizers

**Speakers**

Tim Benton
https://biologicalsciences.leeds.ac.uk/school-of-biology/staff/28/
prof-tim-benton

Daniel Brooks
https://eeb.utoronto.ca/profile/brooks-daniel-r/

Hillary Brown
https://www.postcarbon.org/our-people/hillary-brown/

Martin Rees
https://www.martinrees.uk

Andrew Sheng
http://www.andrewsheng.net

Gert van Santen
https://www.linkedin.com/in/gert-van-santen-54181913/

Alexander Zehnder
https://www.ae-info.org/ae/Member/Zehnder_Alexander

## Invited participants

Gary Dirks
https://sustainability-innovation.asu.edu/person/gary-dirks/

Nik Gowing
https://www.thinkunthink.org/about/our-team/

Carlo Jaeger
https://globalclimateforum.org/portfolio-item/jaeger/

Edward Kirumira
https://sites.wustl.edu/smartafrica/people/edward-kirumira/

Wilhelm Krull
https://thenew.institute/en/who/wilhelm-krull

Tony Mayer
https://www.linkedin.com/in/tony-mayer-080a4017/

Cherie Nursalim
https://www.worldfuturecouncil.org/p/cherie-nursalim/

Peter Robertson
https://www.nyenrode.nl/en/faculty-and-research/visiting-fellows/p/
peter-robertson

A.R.Siders
https://www.drc.udel.edu/people/faculty/siders?uid=siders&Name=
A.R.%20Siders

William Solecki
http://www.geo.hunter.cuny.edu/people/fac/solecki.html

Jan Staman
http://www.stamanconsultancy.com/staman-consultancy

Sander van der Leeuw
https://sustainability-innovation.asu.edu/person/sander-van-der-leeuw/

Coleen Vogel
https://www.wiomsa.org/personnel/coleen-vogel/

Francis Vorhies
https://sowc.alueducation.com/people/dr-francis-vorhies/

Mark Wilson
https://www.linkedin.com/in/mark-wilson-5334517/

## Organizers: Para Limes

Jan W. Vasbinder
jan.w.vasbinder@paralimes.org

Jonathan Y. H. Sim
https://i.am.jyhsim.com

www.ingramcontent.com/pod-product-compliance
Ingram Content Group UK Ltd.
Pitfield, Milton Keynes, MK11 3LW, UK
UKHW020039170726
7214IPUK00036B/302